Contents

The units in this book have been arranged to match the Cambridge Lower Secondary Mathematics curriculum framework. Each unit is colour coded according to the area of the syllabus it covers:

■ Number

■ Geometry & Measure

■ Statistics & Probability

■ Algebra

Cambridge checkp●int

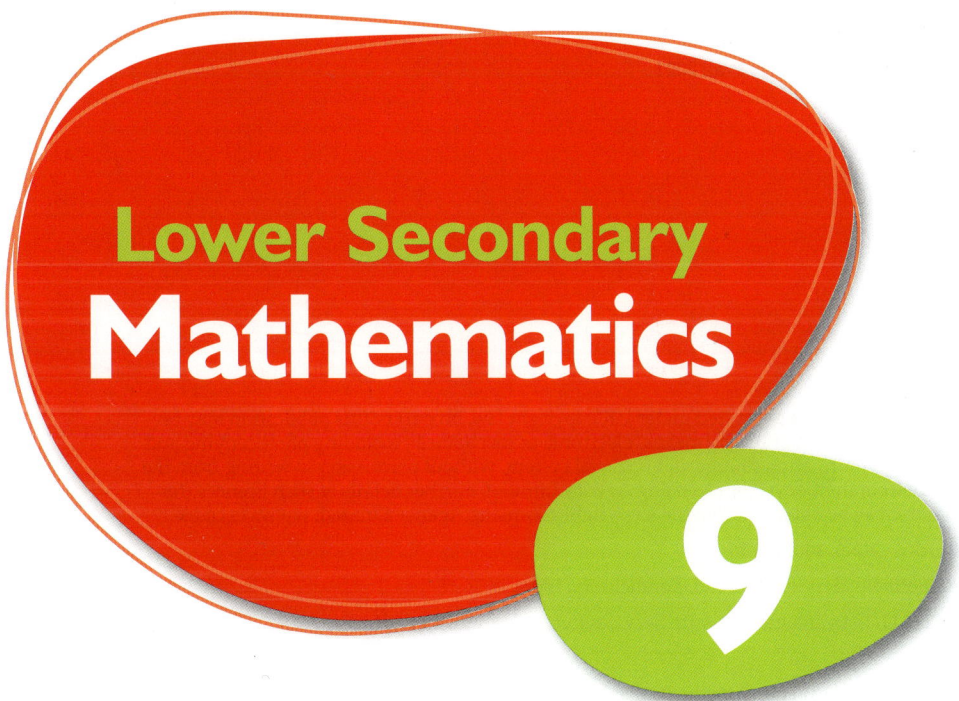

Lower Secondary
Mathematics

9

Ric Pimentel
Frankie Pimentel
Terry Wall

HODDER
EDUCATION
AN HACHETTE UK COMPANY

The Publishers would like to thank the following for permission to reproduce copyright material.

Photo credits
p. vii © Ktsdesign/stock.adobe.com; **p.1** © Chronicle/Alamy Stock Photo; **p.35** royaltystockphoto /123RF.com;
p.40 Courtesy of International Business Machines Corporation, © International Business Machines Corporation;
p.73 © Zlatko Guzmic/stock.adobe.com; **p.125** © Ingram Publishing Limited / Ingram Image Library 500-Sport;
p.167 *t* © North Wind Picture Archives/Alamy Stock Photo, *b* © North Wind Picture Archives/Alamy Stock Photo.

Acknowledgements

Cambridge International copyright material in this publication is reproduced under licence and remains the intellectual property of Cambridge Assessment International Education.

End of section review questions [and sample answers] have been written by the authors. In assessment, the way marks are awarded may be different.

Third-party websites and resources referred to in this publication have not been endorsed by Cambridge Assessment International Education.

Every effort has been made to trace all copyright holders, but if any have been inadvertently overlooked, the Publishers will be pleased to make the necessary arrangements at the first opportunity.

Although every effort has been made to ensure that website addresses are correct at time of going to press, Hodder Education cannot be held responsible for the content of any website mentioned in this book. It is sometimes possible to find a relocated web page by typing in the address of the home page for a website in the URL window of your browser.

Hachette UK's policy is to use papers that are natural, renewable and recyclable products and made from wood grown in well-managed forests and other controlled sources. The logging and manufacturing processes are expected to conform to the environmental regulations of the country of origin.

Orders: please contact Hachette UK Distribution, Hely Hutchinson Centre, Milton Road, Didcot, Oxfordshire, OX11 7HH. Telephone: +44 (0)1235 827827. Email education@hachette.co.uk Lines are open from 9 a.m. to 5 p.m., Monday to Friday. You can also order through our website: www.hoddereducation.com

ISBN: 978 1 3983 0204 4

© Ric Pimentel, Frankie Pimentel and Terry Wall 2021

First published in 2004
Second edition published in 2011
This edition published in 2021 by
Hodder Education,
An Hachette UK Company
Carmelite House
50 Victoria Embankment
London EC4Y 0DZ

www.hoddereducation.co.uk

Impression number 10 9 8 7 6 5 4

Year   2025 2024 2023 2022

Cover photo: © tiverylucky – stock.adobe.com

Illustrations by Integra Software Services Pvt. Ltd, Pondicherry, India

Typeset in Palatino LT Std 11/13 by Integra Software Services Pvt. Ltd, Pondicherry, India

Printed in Dubai

A catalogue record for this title is available from the British Library.

How to use this book

To make your study of Cambridge Checkpoint Lower Secondary Mathematics as rewarding as possible, look out for the following features when you are using the book:

History of mathematics

These sections give some historical background to the material in the section.

These aims show you what you will be covering in the unit.

LET'S TALK
Talk with a partner or a small group to decide your answer when you see this box.

Worked example

These show you how you could approach answering a question.

KEY INFORMATION
These give you hints or pointers to solving a problem or understanding a concept.

These highlight ideas and things to think about.

This book contains lots of activities to help you learn. The questions are divided into levels by difficulty. Green are the introductory questions, amber are more challenging and red are questions to really challenge yourself. Some of the questions will also have symbols beside them to help you answer the questions.

Exercise 15.2

1 Work out the answer to the following calculations. Show your working clearly and simplify your answers where possible.

a $\frac{2}{5}+\frac{1}{6}$

b $\frac{7}{12}+\frac{1}{5}$

c $\frac{9}{14}-\frac{2}{7}$

d $\frac{3}{13}-\frac{3}{26}$

e $\frac{1}{8}+\frac{5}{16}-\frac{5}{24}$

f $\frac{13}{18}-\frac{8}{9}+\frac{1}{6}$

2 Sadiq spends $\frac{1}{5}$ of his earnings on his mortgage. He saves $\frac{2}{7}$ of his earnings. What fraction of his earnings is left?

3 The numerators of two fractions are hidden as shown.

$$\boxed{\frac{\vdots}{8}} + \boxed{\frac{\vdots}{5}} = \frac{23}{40}$$

The sum of the two fractions is $\frac{23}{40}$. Calculate the value of both numerators.

Look out for these symbols:

This green star icon shows the thinking and working mathematically (TWM) questions. This is an important approach to mathematical thinking and learning that has been incorporated throughout this book.

Questions involving TWM differ from the more straightforward traditional question-and-answer style of mathematical learning. Their aim is to encourage you to think more deeply about the problem involved, make connections between different areas of mathematics and articulate your thinking.

This indicates where you will see how to use a calculator to solve a problem.

These questions should be answered without a calculator.

This tells you that content is available as audio. All audio is available to download for free from www.hoddereducation.com/cambridgeextras

There is a link to digital content at the end of each unit if you are using the Boost eBook.

Introduction

This is the third of a series of three books that follow the Cambridge Lower Secondary Mathematics curriculum framework. It has been written by three experienced teachers who have lived or worked either in schools or with teachers in many countries, including England, Spain, Germany, France, Turkey, South Africa, Malaysia and the USA.

Students and teachers studying this course come from a variety of cultures and speak many different languages as well as English. Sometimes cultural and language differences make understanding difficult. However, mathematics is almost a universal language. $28 + 37 = 65$ will be understood in many countries where English is not the first language. A mathematics book written in Japanese will include algebra equations with x and y.

We should also all be very aware that much of the mathematics you will learn in this series of three books was first discovered, and built upon, by mathematicians from China, India, Arabia, Greece and Western European countries.

Most early mathematics was simply game play and problem solving. But later this mathematics was applied to building, engineering and sciences of all kinds. Mathematicians study mathematics because they enjoy it as fun in itself. Mathematicians learn to think and work mathematically.

The DNA molecule discovered by James Watson, Francis Crick, Rosalind Franklin and Raymond Gosling at Cambridge University and King's College, London, has a mathematical structure described as a double-stranded helix. The word helix comes from the Greek meaning 'twisted curve' and looks similar to the shape of a corkscrew.

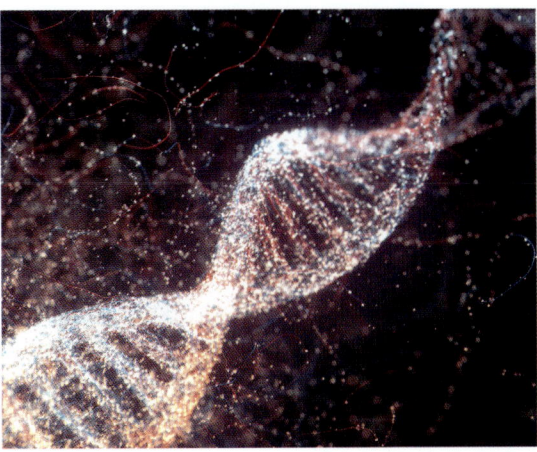

We hope that you will enjoy the work you do and the mathematics you learn in this series of books. Sometimes the ideas will not be easy to understand at first. That should be part of the fun. Ask for help if you need it, but try hard first.

Try to learn by Thinking and Working Mathematically (TWM). This is an important feature of the book. Thinking and Working Mathematically can be divided into the following characteristics:

Thinking and working characteristic	Definition
Specialising	Choosing an example and checking if it satisfies or does not satisfy specific mathematical criteria.
Generalising	Recognising an underlying pattern by identifying many examples that satisfy the same mathematical criteria.
Conjecturing	Forming mathematical questions or ideas.
Convincing	Presenting evidence to justify or challenge a mathematical idea or solution.
Characterising	Identifying and describing the mathematical properties of an object.
Classifying	Organising objects into groups according to their mathematical properties.
Critiquing	Comparing and evaluating mathematical ideas, representations or solutions to identify advantages and disadvantages.
Improving	Refining mathematical ideas or representations to develop a more elegant approach or solution.

Where you see this icon ⭐ it shows you that you will be thinking and working mathematically.

Writing down your thoughts and workings helps to develop your mathematical fluency. By thinking carefully about how you explain your ideas you may, while justifying an answer, be able to make wider generalisations. Discussing different methods with other students will also help you compare and evaluate your mathematical ideas. This will lead to you understand why some methods are more effective than others in given situations. Throughout you should always be forming further mathematical questions and presenting other ideas for thought.

Many students start off by thinking that mathematics is just about answers. Although answers are often important, posing questions is just as important. What is certainly the case is that the more you question and understand, the more you will enjoy mathematics.

The writers of this book believe that with sufficient time and hard work all students can learn to think and work mathematically.

At the end of this series you, the student, should be able to think and work mathematically independently of your teacher. This will be more important as you progress in mathematics.

Ric Pimentel, Frankie Pimentel and Terry Wall, 2021

History of mathematics – The Greeks

"If I were beginning my studies I would follow the advice of Plato and start with mathematics."

Galileo

Arithmetic explores the rules of counting. Algebra looks at the **general** rules of arithmetic. Geometry and trigonometry look at the **general** rules of shapes.

Many of the great Greek mathematicians and scientists lived in, or came from, the Greek islands or from cities like Ephesus or Miletus, which are in present-day Turkey, or from Alexandria in Egypt. The people described below are some of the Greek mathematicians who lived during what was called the 'golden age', from 500 to 300 BCE. You may wish to find out more about them.

The construction of the most famous building in Athens, the Parthenon, required a very high level of mathematics to make it such a work of art.

Thales of Alexandria discovered the 365-day calendar and worked out the dates of eclipses of the Sun and the Moon.

▲ Euclid of Alexandria

Pythagoras of Samos, a Greek island, led a group of mathematicians and worked with geometry. The next leader was Theano, who was the first woman to be a leader in mathematics.

Euclid of Alexandria formed what would now be called a university department. His book became the main textbook in schools and universities for 2000 years.

Apollonius of Perga (Turkey) worked on, and gave names to, the parabola, the hyperbola and the ellipse. (Ask your teacher to draw these.)

Archimedes is accepted today as possibly the greatest mathematician and scientist of all time. However, he was so far ahead of his time that others could not understand his ideas.

1 Indices and standard form

- Use positive, negative and zero indices, and the index laws for multiplication and division.
- Understand the standard form for representing large and small numbers.

Indices

Recap

The **index** of a number refers to the power to which it is raised.

For example, in 5^3 the '3' is the index whilst the '5' is known as the **base number**.

5^3 is equivalent to $5 \times 5 \times 5 = 125$.

The plural of index is **indices**.

In Stage 8 you were introduced to the following laws regarding indices:

- $5^2 \times 5^4 = 5 \times 5 \times 5 \times 5 \times 5 \times 5$
  $\qquad = 5^6$ (i.e. 5^{2+4})

  This can be written in a **general** form as:

  $$a^m \times a^n = a^{m+n}$$

Remember that the base numbers must be the same for this rule to be true.

- $3^6 \div 3^2 = \dfrac{3 \times 3 \times 3 \times 3 \times \cancel{3 \times 3}}{\cancel{3 \times 3}}$
  $\qquad = 3^4$ (i.e. 3^{6-2})

  This can be written in a **general** form as:

  $$a^m \div a^n = a^{m-n}$$

Again, the base numbers must be the same for this rule to be true.

- $(7^2)^3 = (7 \times 7) \times (7 \times 7) \times (7 \times 7)$
  $\qquad = 7^6$ (i.e. $7^{2 \times 3}$)

  This can be written in a **general** form as:

  $$(a^m)^n = a^{mn}$$

> **KEY INFORMATION**
> These are known as the **general laws of indices**.

- The **zero index** means that a number has been raised to the power of 0.
  Any number raised to the power of 0 is equal to 1. For example:
  $$4^0 = 1 \qquad 10^0 = 1 \qquad a^0 = 1$$
  This can be explained by applying the laws of indices.
  $$a^m \div a^n = a^{m-n}$$
  Therefore
  $$a^m \div a^m = a^{m-m}$$
  $$= a^0$$
  But
  $$a^m \div a^m = 1$$
  Therefore
  $$a^0 = 1$$

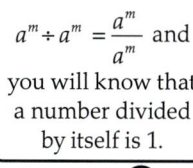

$a^m \div a^m = \dfrac{a^m}{a^m}$ and you will know that a number divided by itself is 1.

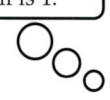

Negative indices

A **negative index** means that a number has been raised to a negative power, for example 4^{-3}.

There are other ways of writing negative powers, as proved below.

For example, 4^{-3} can also be written as 4^{0-3} which, as was shown above, can be written as:
$$4^0 \div 4^3 \text{ or } \dfrac{4^0}{4^3}$$

As $4^0 = 1$ we can deduce that $\dfrac{4^0}{4^3} = \dfrac{1}{4^3}$

Therefore $4^{-3} = \dfrac{1}{4^3}$

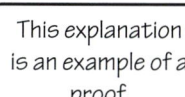

This explanation is an example of a proof.

LET'S TALK
Why is this explanation considered a proof?

So in **general**:
$$\boldsymbol{a^{-m} = \dfrac{1}{a^m}}$$

This can also be demonstrated using numbers:
$$4^2 = 4 \times 4$$
$$4^1 = 4 \quad \Big) \div 4$$
$$4^0 = 1 \quad \Big) \div 4$$
$$4^{-1} = \dfrac{1}{4} \quad \Big) \div 4$$
$$4^{-2} = \dfrac{1}{4^2} \quad \Big) \div 4$$

Worked examples

1   Write 4^{-2} as a fraction.

$$4^{-2} = \frac{1}{4^2}$$

$$= \frac{1}{16}$$

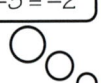

When multiplying, add the indices.
$-4 + 5 + -3 = -2$

2   Simplify the following: $7^{-4} \times 7^5 \times 7^{-3}$

Give your answer as a fraction.

$$7^{-4} \times 7^5 \times 7^{-3} = 7^{-2}$$

$$= \frac{1}{7^2}$$

$$= \frac{1}{49}$$

3   Work out the following: $3^4 \times 3^{-6}$

Give your answer in positive index form.

$3^4 \times 3^{-6} = 3^{-2}$ *Although this is correct, the index is not a positive number.*

$$3^{-2} = \frac{1}{3^2}$$

Exercise 1.1

Without using a calculator, write each of these as an integer or a fraction.

1   a 2^{-3} c 3^{-2} e 2^{-5}
    b 5^{-2} d 4^{-3}

2   a 4×4^{-1} c 10×10^{-2} e 1000×10^{-2}
    b 9×3^{-2} d 500×10^{-3}

3   a 27×3^{-2} c 64×2^{-3} e 36×6^{-3}
    b 16×2^{-3} d 4×2^{-3}

4   a 27×3^{-4} c 4×2^{-3} e 6×6^{-3}
    b 16×2^{-5} d 4×2^{-4}

5   a $2^{-3} \times 2^5$ c $2^7 \times 2^{-5}$ e $2^{-3} \times 2^8 \times 2^{-4}$
    b $5^{-2} \times 5^3$ d $4^3 \times 4^{-5}$

6   Using indices, find the value of each of these.
    Give your answers in positive index form.

    a $5^2 \times 5^{-3}$ d $10^3 \div 10^8$ g $10^4 \times 10^{-3} \div 10^6$
    b $3^5 \times 3^4$ e $5^4 \div 5^5$ h $3^{-2} \times 3 \div 3^7$
    c $6^2 \times 6^{-3} \times 6$ f $8^3 \div 8^3 \times 8^{-2}$ i $4^3 \times 4^{-5} \div 4^8$

7 12 cards are shown below.

| $\dfrac{1}{4}$ | $\dfrac{1}{25}$ | $\dfrac{1}{8}$ | $\dfrac{1}{27}$ | $\dfrac{1}{36}$ | $\dfrac{1}{1000}$ |

| 2^{-3} | 6^{-2} | 10^{-3} | 2^{-2} | 5^{-2} | 3^{-3} |

Pair each fraction with its corresponding number written using indices.

8 A square has a side length of $\dfrac{1}{3}$ cm as shown.

Show that its area is 3^{-2} cm².

$\dfrac{1}{3}$ cm

$\dfrac{1}{3}$ cm

9 A cube has a side length of $\dfrac{1}{4}$ cm.

a   Show that its volume is 2^{-6} cm³.
b   Show that its total surface area is $\dfrac{3}{2^{3}}$ cm².

10 A right-angled triangle has dimensions as shown.
Calculate its area, giving the answer as a fraction with a positive index.

2^{-1} cm

4^{-1} cm

11 The cuboid below has dimensions as shown.

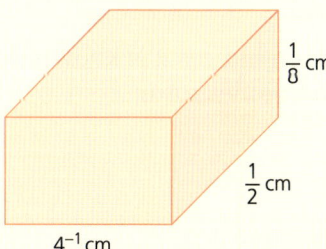

$\dfrac{1}{8}$ cm

$\dfrac{1}{2}$ cm

4^{-1} cm

a   Show that the volume is 2^{-6} cm³.
b   Calculate the total surface area of the cuboid and give your answer in the form $a \times 2^{b}$, where a and b are integers.

Standard form

You will be quite used to describing a distance between two towns as 50 km, say, or the mass of a newborn baby as being 3.2 kg. These numbers have not got many digits and so are easy to read and understand.

However, there are times when just writing the number in full does not make it clearer, in fact it makes it more difficult to understand.

The distance between Earth and the Sun is 149 000 000 km (to three significant figures).

To read this number you will probably have counted the number of zeros and, as there are six of them, have come to the conclusion that the distance is 149 million km.

However, if the question is 'How many atoms are there in a grain of sand?', a possible answer is 330 000 000 000 000 000 000.

This is a very large number. How is it pronounced?

How many zeros are there?

The number is in fact 330 quintillion.

Investigate the names of other large numbers.

The names for numbers bigger than 'billions', though, are not generally well known so a better system for describing large numbers is used.

From the work on indices it is known that the following numbers can be written in index form:

$$100 = 10^2$$

$$1000 = 10^3$$

$$10\,000 = 10^4$$

$$100\,000 = 10^5$$

$$1\,000\,000 = 10^6$$

etc.

The distance between Earth and the Sun can therefore be written as 149×10^6 km. But it could also be written in many other ways, such as 14.9×10^7 km or $14\,900 \times 10^4$ km.

The correct way of writing it is as 1.49×10^8 km. When a number is written like this it is called writing the number in **standard form**.

In general, a number in standard form is written as $a \times 10^n$ where a must be a number greater than or equal to 1 but smaller than 10. Also, n must be an integer value.

KEY INFORMATION

The value of a can be written as an inequality as $1 \leqslant a < 10$.

Worked example

The distance between London and New York is 5585 km.

a Write down the distance in standard form.

$$5585 = 5.585 \times 10^3 \text{ km}$$

b Write the distance in metres, in standard form and correct to two significant figures.

Converting km to m:

$$5585 \text{ km} = 5\,585\,000 \text{ m}$$

To two significant figures this is written as $5\,600\,000$ m.

In standard form it is 5.6×10^6 m.

Remember that with 2 s.f. you look at the third digit. If it is 5 or more, the number is rounded up.

Exercise 1.2

1 a **Classify** these numbers into two groups: numbers in standard form and numbers not in standard form.

4.2×10^5 $\qquad$ 3.678×10^2 $\qquad$ 12×10^3 $\qquad$ 463×10^5

b For those that are not written in standard form:

 i) explain why they are not

 ii) rewrite them in standard form.

2 Write each of the following numbers in standard form to two significant figures.

 a 7000

 b 42 100 000 000

 c 55 500 000

 d 120

3 The population of Wales in the United Kingdom is 3.1×10^6.

 The ratio of the population of Wales to that of China is approximately 1 : 450.

 Calculate the population of China, giving your answer in standard form.

4 Use a calculator to calculate the following. Leave your answers in standard form.

 a $(3.7 \times 10^5) + (4.2 \times 10^4)$

 b $(8.7 \times 10^9) - (6.8 \times 10^4)$

 c $(1.5 \times 10^5) \times (1.8 \times 10^2)$

 d $(1.9 \times 10^9) \div (4.2 \times 10^6)$

5 The mass of a grain of rice is on average 0.029 grams.

 a Calculate the approximate number of grains in a 1 kg bag of rice.

 b Write your answer to part (a) in standard form to 2 s.f.

6 The following table shows the average distance of each planet in our solar system from Earth.

Planet	Distance from Earth (km)
Mercury	91.7 million
Venus	41.4 million
Mars	78.3 million
Jupiter	628.7 million
Saturn	1275.0 million
Uranus	2724.0 million
Neptune	4351.4 million

 a Write each of the distances in standard form correct to 3 s.f.

 b A space ship travels at an average speed of 2.8×10^4 km per hour.

  i) How many days would it take to travel from Earth to Mars?

  ii) How many years would it take to travel from Earth to Neptune?

If you have a bicycle, work out how many times its wheel would rotate if you took up the challenge of cycling around the equator.

7 The circumference of the Earth around the equator is 40 075 km. As a challenge and to raise money for charity, a cyclist decides to ride that distance in the gym. The wheel of her bicycle has a radius of 38 cm.

Calculate the number of times the wheel will rotate during the challenge. Give your answer in standard form correct to 3 s.f.

A chessboard is made up of 64 squares.

8 An Indian legend tells the story of a king who lost a game of chess. When he lost, the king asked his opponent what he wanted as a reward.

His opponent said he wanted the king to place one grain of rice on the first square of the chessboard, two on the second square, four on the third square and so on, doubling the amount of rice on each square.

LET'S TALK

How many grains of rice in total are needed for the whole board?

This is very challenging and is a project in itself!

a What is the total number of grains of rice used for the first row of 8 squares?
b What is the total number of grains of rice used for the first two rows of the board?
c How many grains of rice would be placed on the 64th square of the board? Give your answer in standard form to 3 s.f.

LET'S TALK

If the rice could be laid down to a depth of 1 m, can you estimate the area it would cover?

Small numbers

So far we have only looked at very large numbers, but reading and understanding very small numbers can also be difficult.

LET'S TALK

If builders' sand comes in 25 kg packs, how many grains of sand are in one pack?

The mass of a grain of sand obviously varies a lot depending on the type of sand. However, assume that an engineer has worked out the mass of an average grain of sand to be 0.000 004 5 kg.

This is quite difficult to visualise as an amount as it is so small.

Changing the mass into grams makes it slightly easier (i.e. 0.004 5 g) but it is not always possible to do this.

Standard form can also be used to write very small numbers.

Look at the number pattern below:

$$2 \times 10^3 = 2000$$
$$2 \times 10^2 = 200$$
$$2 \times 10^1 = 20$$
$$2 \times 10^0 = 2$$
$$2 \times 10^{-1} = 0.2$$
$$2 \times 10^{-2} = 0.02$$
$$2 \times 10^{-3} = 0.002$$

As can be seen, as the power of 10 decreases by 1, the answer is divided by 10. This is because standard form involves the use of indices.

KEY INFORMATION

Remember with indices:
$$10^0 = 1$$
$$10^1 = 10$$

Therefore 2×10^{-1} is the same as $2 \times \dfrac{1}{10} = 0.2$.

Similarly, 2×10^{-2} is the same as $2 \times \dfrac{1}{10^2} = 2 \times \dfrac{1}{100} = 0.02$.

To write a very small number in standard form the same rule applies, i.e. that it is written as $a \times 10^n$ where a must be a number equal to or greater than or equal to 1 but smaller than 10 and n must be an integer value.

Worked examples

1 Write 0.000 056 in standard form.

$$0.000\,056 = 5.6 \times 10^{-5}$$

2 Using a calculator work out the following calculation:

$$(3 \times 10^{-3}) \times (4 \times 10^{-4})$$

Give your answer

a in standard form   b in full.

$$(3 \times 10^{-3}) \times (4 \times 10^{-4}) = 1.2 \times 10^{-6}$$ $0.000\,001\,2$

Exercise 1.3

1 Write the following numbers in standard form.
   a 0.000 64
   b 0.000 000 576
   c 0.21
   d 0.1

2 Write the answer to the following calculations
   i) in standard form
   ii) in full.
   a $(7 \times 10^{-3}) \times (6 \times 10^{-5})$
   b $(1.2 \times 10^{-2}) \div (2 \times 10^{-4})$
   c $(1.6 \times 10^{-4}) + (3.2 \times 10^{-3})$
   d $(4.5 \times 10^{-3}) - (6.2 \times 10^{-4})$

3 Using your knowledge of indices and standard form, explain why the following calculations are correct.
   a $(2 \times 10^{-4}) \times (3 \times 10^{2}) = 6 \times 10^{-2}$
   b $(4.1 \times 10^{-6}) \times (2 \times 10^{-2}) = 8.2 \times 10^{-8}$
   c $(3 \times 10^{-5}) + (4 \times 10^{-5}) = 7 \times 10^{-5}$
   d $(8 \times 10^{-3}) - (6 \times 10^{-3}) = 2 \times 10^{-3}$

4 Identify the smaller number in each of the following pairs.
   a 6.7×10^{-5} 8.0×10^{-6}
   b 0.000 81 9.0×10^{-4}
   c 0.000 000 09 1.0×10^{-7}

5 The following two lists show calculations involving numbers in standard form on the left-hand side and answers on the right-hand side.

$(4 \times 10^{-4}) \times (3 \times 10^{2})$ 1.2×10^{-4}

$(6 \times 10^{-8}) \times (2 \times 10^{-3})$ 1.2×10^{-10}

$(5.2 \times 10^{-4}) - (4 \times 10^{-4})$ 1.2×10^{-9}

$(6 \times 10^{-6}) \div (5 \times 10^{3})$ 1.2×10^{-1}

  a   Match each calculation to its correct answer.

  b   Justify your selections in part (a).

6 The diameter of human hair is approximately $0.07\,\text{mm}$.
Write the diameter of human hair in metres in standard form.

7 A box of paper contains 4 packs of paper, with 500 sheets in each pack.
The total height of the box is $26\,\text{cm}$.
Calculate the thickness of one sheet of paper in cm, giving your answer in standard form.

8 In a physics book, the masses of an electron, a proton and a neutron are listed as follows:

Mass of an electron $= 9.109 \times 10^{-31}\,\text{kg}$
Mass of a proton $= 1.673 \times 10^{-27}\,\text{kg}$
Mass of a neutron $= 1.675 \times 10^{-27}\,\text{kg}$

Write these masses in order of size starting with the smallest.

KEY INFORMATION

Electrons, protons and neutrons are parts of an atom.

Now you have completed Unit 1, you may like to try the Unit 1 online knowledge test if you are using the Boost eBook.

Pythagoras' theorem

● Know and use Pythagoras' theorem.

Exercise 2.1

1
- In the centre of a piece of paper, draw a right-angled triangle.
- Off each of the sides of the triangle, construct a square, as shown in the diagram.

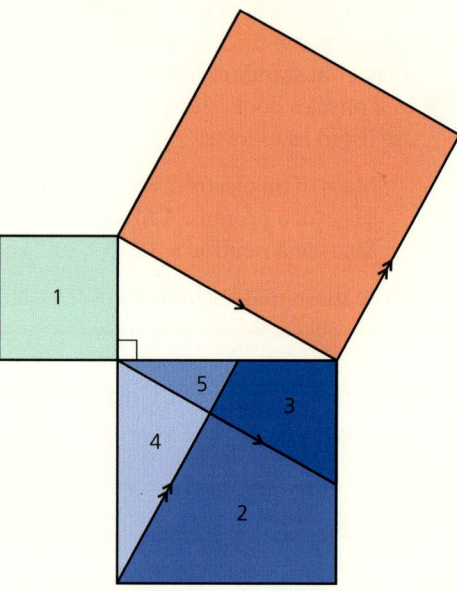

- Divide up one of the smaller squares as shown, making sure the dividing lines run parallel to the sides of the largest square.
- Cut out the shapes numbered 1, 2, 3, 4 and 5, and try to arrange them on top of the largest square so that they fit without any gaps.
- What **conjecture** can you make about the areas of the three squares? Can you find a **general** rule?

KEY INFORMATION

This is an example of a geometric proof.

This dissection to prove Pythagoras' theorem was constructed by Thabit ibn Quarra in Baghdad in 836 BCE.

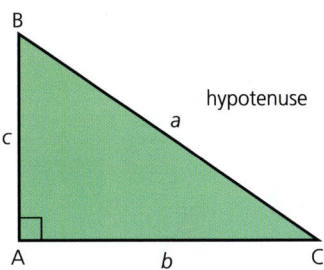

Pythagoras' theorem states the relationship between the lengths of the three sides of a right-angled triangle:

$$a^2 = b^2 + c^2$$

In words, the theorem states that the square of the length of the hypotenuse is equal to the sum of the squares of the other two sides.

Worked examples

1 Calculate the length of the side marked a in this diagram.

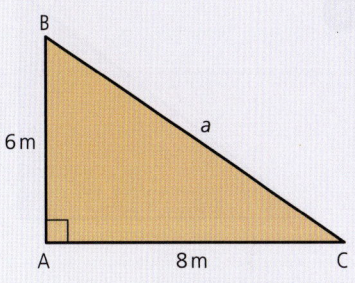

Using Pythagoras' theorem:

$$a^2 = b^2 + c^2$$
$$a^2 = 8^2 + 6^2$$
$$a^2 = 64 + 36 = 100$$
$$a = \sqrt{100} = 10$$

The length of side a is 10 m.

2 Calculate the length of the side marked b in this diagram.

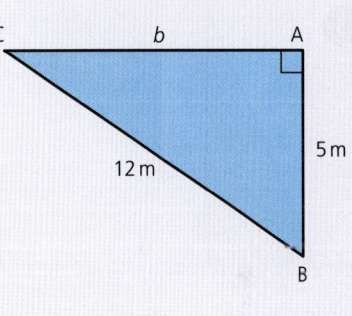

Pythagoras' theorem states that:

$$a^2 = b^2 + c^2$$

Rearranging to make b^2 the subject gives:

$$b^2 = a^2 - c^2$$
$$b^2 = 12^2 - 5^2$$
$$b^2 = 144 - 25 = 119$$
$$b = \sqrt{119}$$

The length of side b is 10.9 m (to one decimal place).

Exercise 2.2

Use Pythagoras' theorem to calculate the length of the hypotenuse in each of these right-angled triangles.

1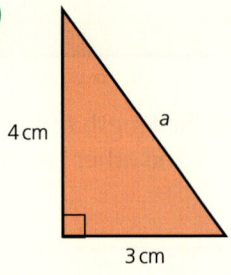

4 cm
a
3 cm

4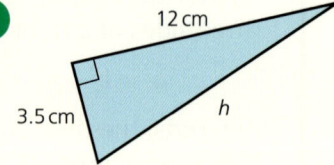

12 cm
3.5 cm
h

2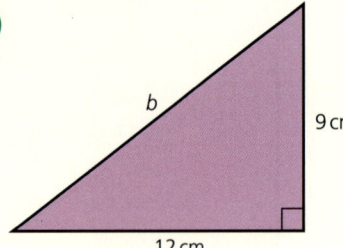

b
9 cm
12 cm

5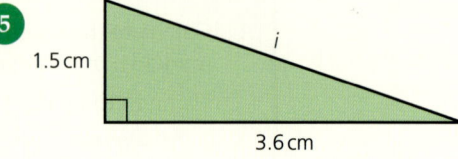

1.5 cm
i
3.6 cm

3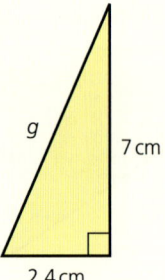

g
7 cm
2.4 cm

6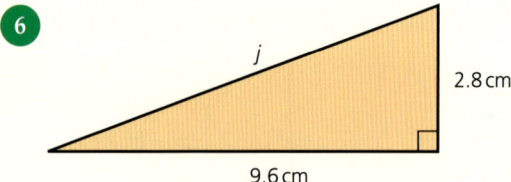

j
2.8 cm
9.6 cm

7 Pablo and Amala set off from a point A.
Pablo heads to a point B before turning and heading towards point C as shown. Amala goes directly from A to C.
The angle ABC is 90°.
Calculate, in metres, how much less far Amala has to walk compared with Pablo.

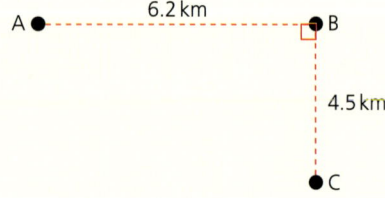

A
6.2 km
B
4.5 km
C

8 Four triangles A, B, C and D are shown below.

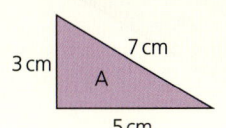
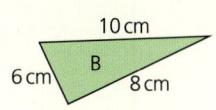
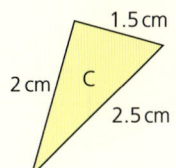
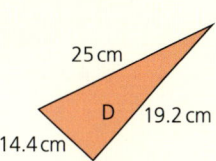

a **Classify** the triangles into right-angled triangles and non-right-angled triangles.
b Justify your choices in part (a).
c For the triangle(s) that are not right-angled, change the length of the hypotenuse in order to make it a right-angled triangle.

9 A field in the shape of a pentagon is shown to the right.
The farmer wants to build a fence around the edge of the field.
The fencing costs $85 per metre.
Calculate the total cost of fencing the field.

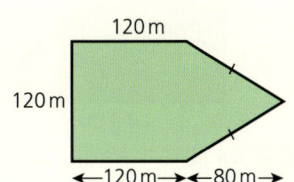

10 The diagram shows a rectangle placed inside a semicircle.
Points P and Q lie on the curved edge of the semicircle whilst the edge SR of the rectangle lies along the straight edge of the semicircle.
The dimensions of the rectangle are 7.5 × 20 cm as shown.
Calculate the curved length of the semicircle. Give your answer correct to 1 d.p.

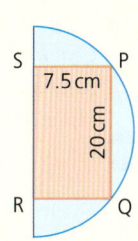

Exercise 2.3

Use Pythagoras' theorem to calculate the length of the unknown side in each of these diagrams.

1

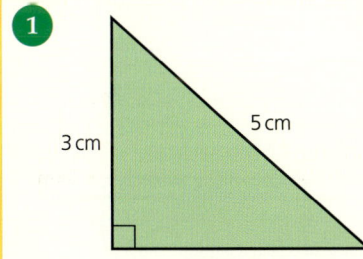

2

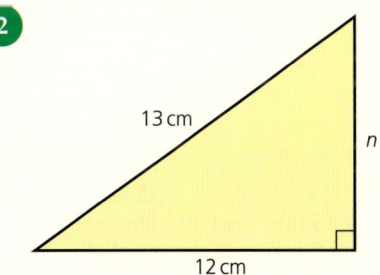

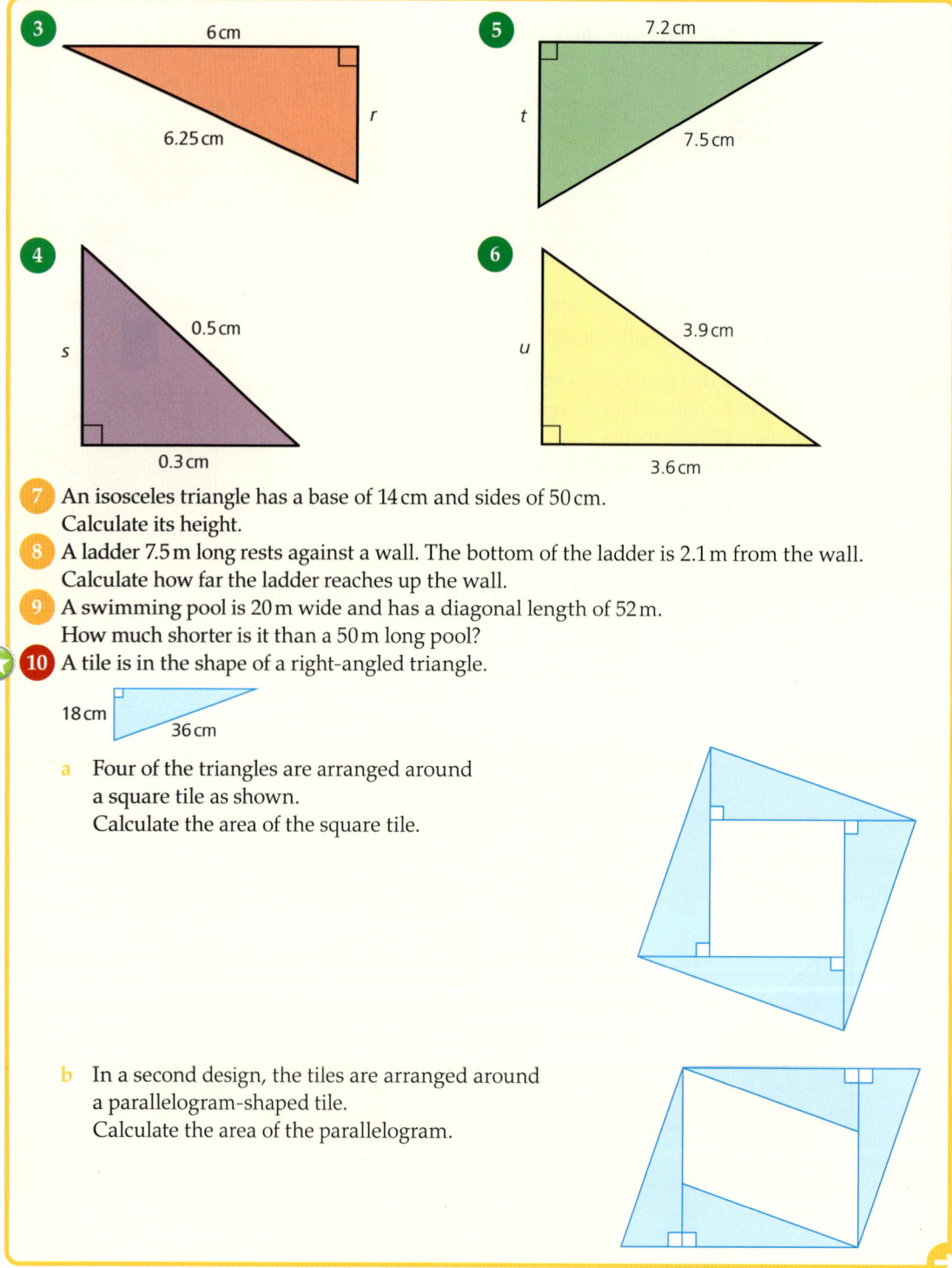

3 6 cm 6.25 cm r

5 7.2 cm 7.5 cm t

4 0.5 cm s 0.3 cm

6 3.9 cm u 3.6 cm

7 An isosceles triangle has a base of 14 cm and sides of 50 cm.
Calculate its height.

8 A ladder 7.5 m long rests against a wall. The bottom of the ladder is 2.1 m from the wall.
Calculate how far the ladder reaches up the wall.

9 A swimming pool is 20 m wide and has a diagonal length of 52 m.
How much shorter is it than a 50 m long pool?

10 A tile is in the shape of a right-angled triangle.

18 cm 36 cm

a   Four of the triangles are arranged around
a square tile as shown.
Calculate the area of the square tile.

b   In a second design, the tiles are arranged around
a parallelogram-shaped tile.
Calculate the area of the parallelogram.

11 The right-angled triangle below has dimensions as shown.

Calculate the value of $\dfrac{q}{p}$.

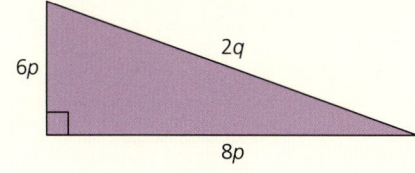

Worked example

Find the lengths (in centimetres) of the sides marked p and q in this diagram.

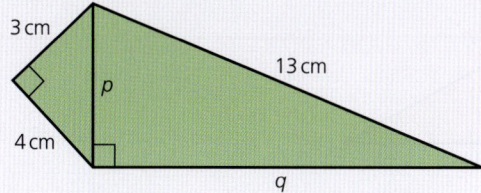

Using Pythagoras' theorem:

$$p^2 = 4^2 + 3^2$$
$$p^2 = 16 + 9$$
$$p^2 = 25$$
$$p = 5$$

Using Pythagoras' theorem again:

$$13^2 = p^2 + q^2$$
$$13^2 = 5^2 + q^2$$
$$13^2 - 5^2 = q^2$$
$$169 - 25 = q^2$$
$$144 = q^2$$
$$q = 12$$

Therefore $p = 5$ cm and $q = 12$ cm.

Exercise 2.4

Find the length of the unknown side in each of these diagrams.

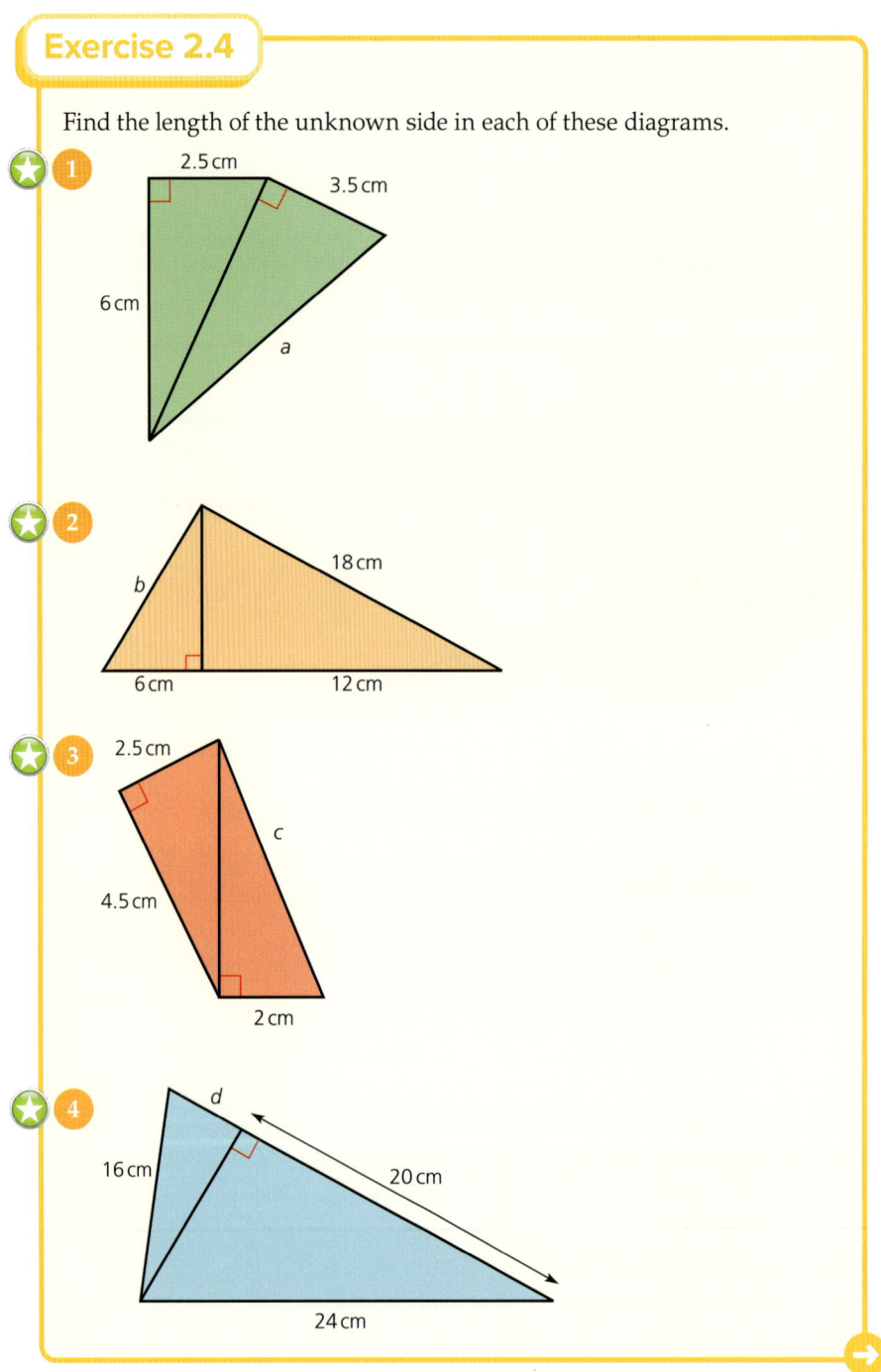

1 2.5 cm, 3.5 cm, 6 cm, a

2 b, 18 cm, 6 cm, 12 cm

3 2.5 cm, c, 4.5 cm, 2 cm

4 d, 16 cm, 20 cm, 24 cm

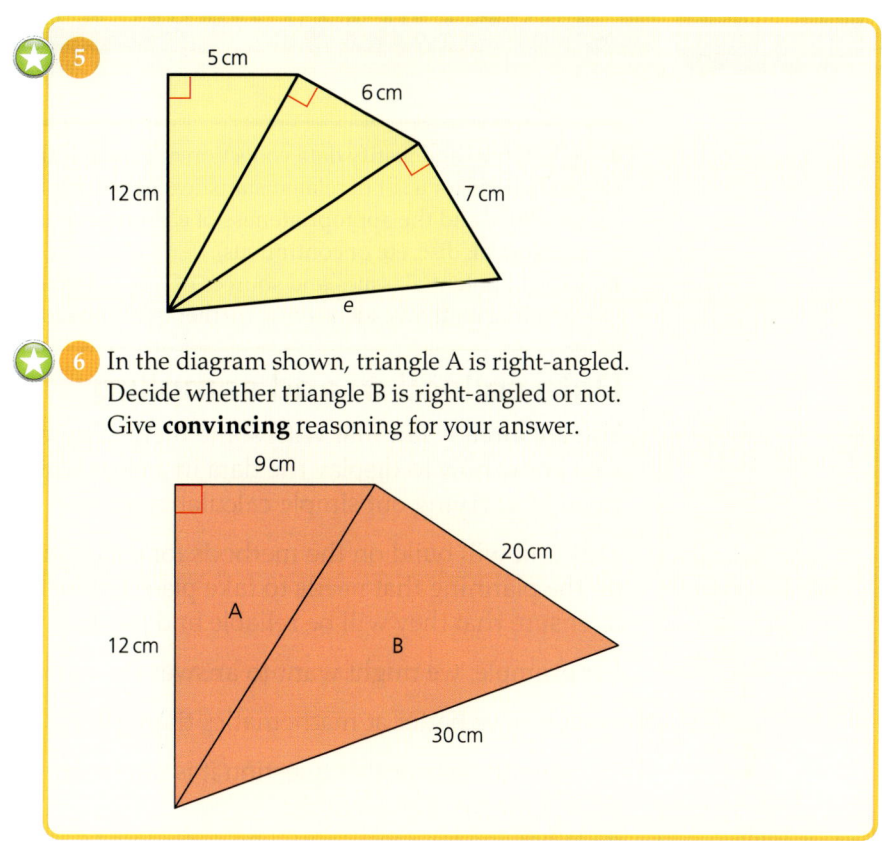

5 5 cm

6 cm

12 cm

7 cm

e

6 In the diagram shown, triangle A is right-angled.
Decide whether triangle B is right-angled or not.
Give **convincing** reasoning for your answer.

9 cm

20 cm

A

12 cm

B

30 cm

Now you have completed Unit 2, you may like to try the Unit 2
online knowledge test if you are using the Boost eBook.

3 Data collection and sampling

- Select, trial and justify data collection and sampling methods to investigate predictions for a set of related statistical questions, considering what data to collect, and the appropriateness of each type (qualitative or quantitative; categorical, discrete or continuous).
- Explain potential issues and sources of bias with data collection and sampling methods, identifying further questions to ask.

Data collection and sampling

This next section can be undertaken as a mini project. It describes the main things to consider when collecting data.

You are already familiar with some methods for collecting data. You also know how to display the data in a clear way and how to analyse them by carrying out simple calculations.

This unit will build on the methods for collecting data, in particular on the planning that needs to take place before any data are collected to ensure that they will be reliable and produce useful information.

For example, we might want to answer the following question:

‘Are boys better at mathematics than girls?’

To begin to answer this question it is important to be precise about what it means. For example:
- What age (or ages) should be considered?
- What is meant by ‘better at mathematics’?

LET'S TALK

How can the question be made more precise?

Once the question has been decided on, it is important to think about where your data will come from. For example:
- You may collect the data yourself. These are called **primary data**.
- You may use data collected by your teachers or from the internet. These are called **secondary data**.

LET'S TALK

What primary data could be collected to answer the question ‘Are boys better at mathematics than girls?’ Discuss your suggestions.

What types of secondary data could be collected? Discuss your suggestions.

Once the sources of data are decided on, you need to decide how the data are going to be collected.
- Primary data are usually collected by questionnaire or by interview and are examples of **qualitative data**.

● Secondary data will often be in number form. Numerical data are also known as **quantitative data**.
How will these data be collected?

LET'S TALK

Discuss how both the primary and secondary data could be collected.

If you have decided on using a questionnaire for the primary data, design a first draft for the questionnaire.

If the secondary data are in number form, design tables (or a spreadsheet) in which to collect the data.

A good sample must be representative of the whole population and therefore give similar results to the whole population.

You also need to make a decision about how much data are to be collected. For this investigation, it would clearly not be possible to collect data from every student in the country or region, or even in the whole school. You will need to decide on a **sample size**, in other words how many people will be included in the investigation. It is also important to discuss how you are going to select people within the sample. For example:

● If you are looking at students from Years 7–11, will you choose the same number from each year?
● Will you choose the same number of boys and girls?
● Will the students you choose be your friends or will they be picked randomly?

KEY INFORMATION

For a sample to be representative it must avoid **bias**. Why might picking just your friends be an example of bias?

LET'S TALK

Discuss the sample sizes you will use for both the primary and the secondary data.

Discuss how the samples are to be chosen to ensure that students are picked randomly.

You also need to decide how accurate the data collection needs to be. This may seem an odd consideration: it seems obvious that the data collection should be as accurate as possible. However, this is not necessarily the case. For example:

LET'S TALK

Decide on the level of accuracy needed for your data.

● If exam results are being collected and the data are in percentages, is it sensible to count the number of students who got 1%, 2%, 3%, … ?
● Would it be better to group the data and count the number of students getting 1–10%, 11–20%, 21–30%, … ?

Once all of these points have been carefully considered it is likely that the data collected will be appropriate and that any conclusions made will be valid. However, it is always possible that something has been overlooked or that the data collection does not work out exactly as planned. Before starting the actual data collection, it is always sensible to do a trial of the whole process. For example:

- If you are using a questionnaire, test it with a few people first to make sure the questions are clear.
- If you are organising your data in a spreadsheet, make sure that it works with a small amount of data first.
- You may decide that the sample size is too big or too small.

Once you have carried out a trial on the different stages of your data collection process, you need to **critique** your trial and make any **improvements** that you think are necessary.

LET'S TALK

What **improvements** have you made, and why did you decide to make them?

Exercise 3.1

Choose one of the suggested questions 1–5 below as an area for investigation and data collection. Then:

a   plan your data collection following the steps outlined in this unit.

b   carry out the data collection in full.

c   present your findings in a clear way.

You need to use:

- **characterising** skills to identify and describe properties of the data that you collect
- **classifying** skills to organise the data into groups
- **critiquing** skills to identify any problems or strengths with your trial
- **improving** skills to refine your investigation.

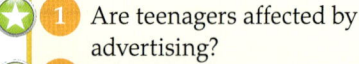
1   Are teenagers affected by advertising?

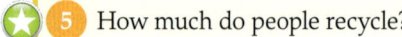
2   Do teenagers have a good diet?

3   Is mathematics more important than art?

4   How do people's spending habits change as they get older?

5   How much do people recycle?

KEY INFORMATION

There are many different ways to present findings. These can be in table form, or in a range of different graph forms.

KEY INFORMATION

There are different types of data. These include **discrete, continuous, qualitative, quantitative** and **categorical** data. The decisions you take will often be affected by the type of data you are collecting.

By completing a data collection exercise thoroughly you will have noticed that it is not an exact science. The planning decisions you make will affect what data are collected, how they are collected and how accurate they are. Therefore, it is important to plan things thoroughly and be able to justify any decisions you make.

Area and circumference of a circle

- Know and use the formulae for the area and circumference of a circle.
- Estimate and calculate areas of compound 2D shapes made from rectangles, triangles and circles.

Recap

Circumference of a circle

You already know from Stage 8 that the **circumference** of any circle is given by the formula:

Circumference = $\pi \times$ **diameter** or $C = \pi D$

As the **diameter** is twice the **radius,** the **circumference** of a circle can also be given as:

Circumference = $\pi \times 2 \times$ **radius** or $C = 2\pi r$

Pi (π) is not an exact number; it has an infinite number of decimal places.

To two decimal places, $\pi = 3.14$

To 14 decimal places, $\pi = 3.141\,592\,653\,589\,79$

Scientific calculators have a $\boxed{\pi}$ key. Check to see how many decimal places your calculator gives pi to.

Worked example

Calculate the perimeter of the compound shape below.

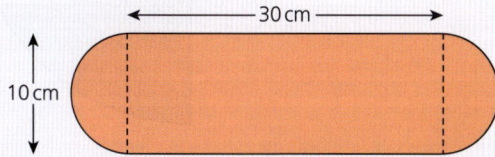

This compound shape can be divided into a rectangle and two semicircles.

Perimeter of outside edges of the rectangle = $2 \times 30 = 60\,$cm

The two semicircles if combined would form a complete circle, therefore the perimeter of the two semicircles is equal to the circumference of the full circle.

Circumference = $\pi \times 10 = 31.4\,$cm (to 1 d.p.)

Therefore, total perimeter of compound shape is $60 + 31.4 = 91.4\,$cm.

Exercise 4.1

1 Calculate the circumference of each of these circles. The diameter of each circle has been given. Give your answers correct to two decimal places.

a

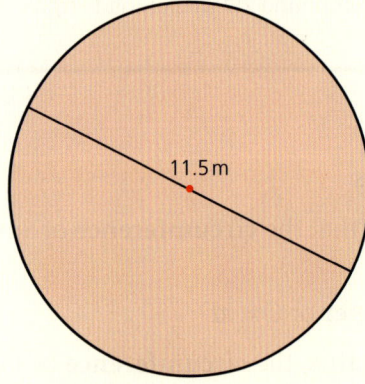

11.5 m

b

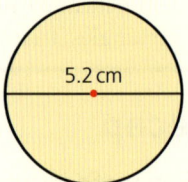

5.2 cm

2 Calculate the circumference of each of these circles. The radius of each circle has been given. Give your answers correct to two decimal places.

a

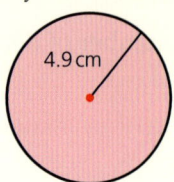

4.9 cm

b

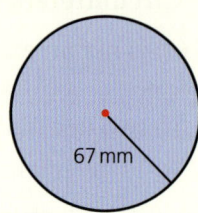

67 mm

3 Calculate the perimeter of each of these shapes. Give your answers to a suitable degree of accuracy.

a

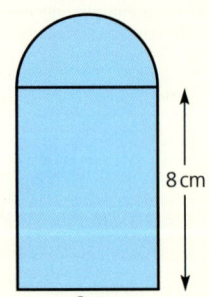

8 cm

6 cm

b

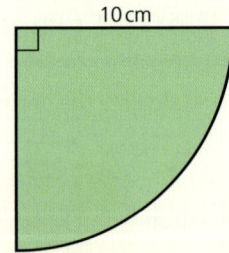

10 cm

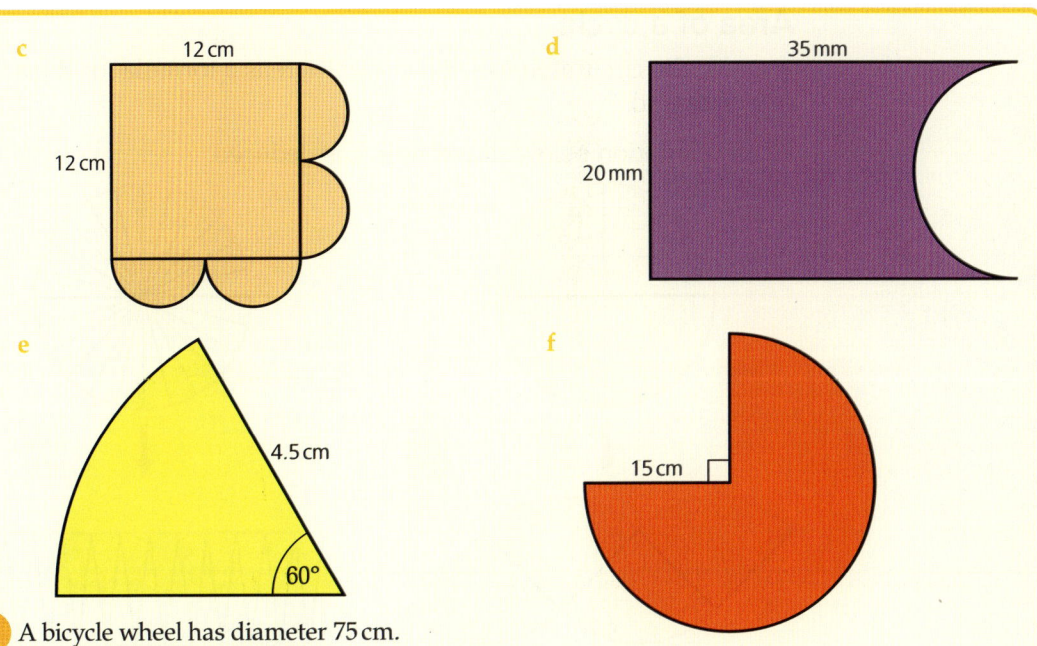

c

12 cm

12 cm

d

35 mm

20 mm

e

4.5 cm

60°

f

15 cm

4 A bicycle wheel has diameter 75 cm.
  a Calculate the length of its circumference to one decimal place.
  b How many times will the wheel rotate if a girl rides the bicycle for 1 km?
    Give your answer correct to the nearest whole number.

5 A circular hole of radius 49 m is cut from a circular metal disc of radius 50 cm, to produce a ring as shown.

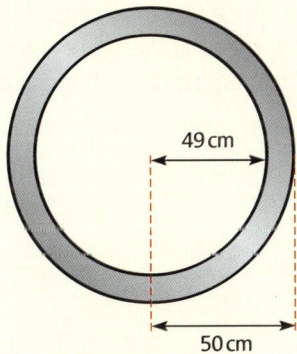

49 cm

50 cm

How much longer is the outer circumference of the ring compared with the inner circumference?
Give your answer correct to one decimal place.

6 A school has a circular athletics track. The inner track has radius 40 m and the middle track has radius 46 m. Two friends decide to race each other. One runs on the inside track, whilst the other runs on the middle track. If both of them run a complete lap of the track, calculate the difference in the distances they have to run. Give your answer correct to two decimal places.

Area of a circle

As with the circumference of a circle, there is a formula for the area of a circle too.

The explanation below shows how it is deduced.

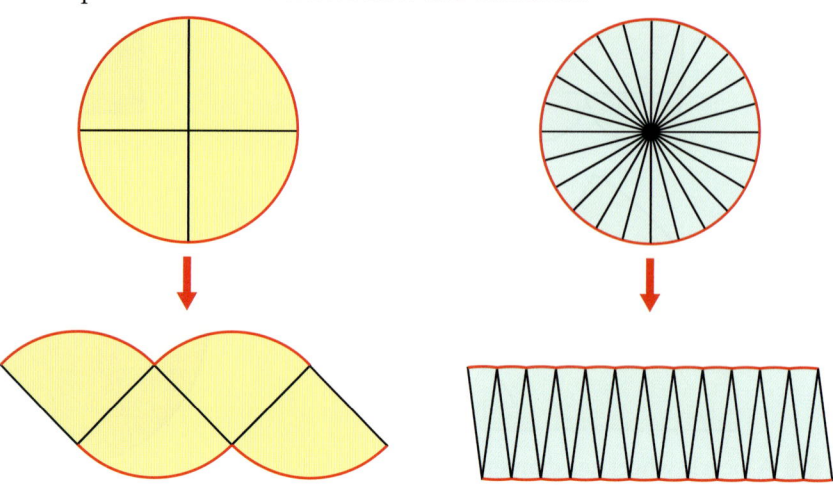

Both of the diagrams show a circle split into sectors. In the first diagram, the circle has been divided into four sectors, whilst in the second diagram the circle has been split into 24 sectors. The sectors are then rearranged to form another shape. As can be seen, the more the number of sectors increases, then the more the rearranged shape looks like a rectangle. In theory, therefore, if the circle were divided into an infinite number of sectors, then the rearranged shape would be a rectangle. This is shown below.

LET'S TALK

Discuss how the dimensions of the rectangle relate to those of the circle.

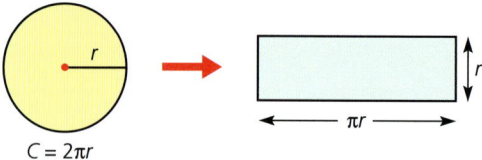

Then:

Area of the circle = area of the rectangle

$$= \text{length} \times \text{width}$$

$$= \pi \times r \times r$$

$$= \pi r^2$$

Area of a circle $= \pi r^2$

Worked example

A circle of diameter 20 cm has two circles of diameter 10 cm cut from it as shown.

Calculate the shaded area.

Area of large circle $= \pi r^2$

$\qquad = \pi \times 10^2$

$\qquad = 314\,\text{cm}^2$ (to 3 s.f.)

Area of a small circle $= \pi \times 5^2$

$\qquad = 78.5\,\text{cm}^2$

Therefore shaded area $= 314 - 2 \times 78.5$

$\qquad = 157\,\text{cm}^2$

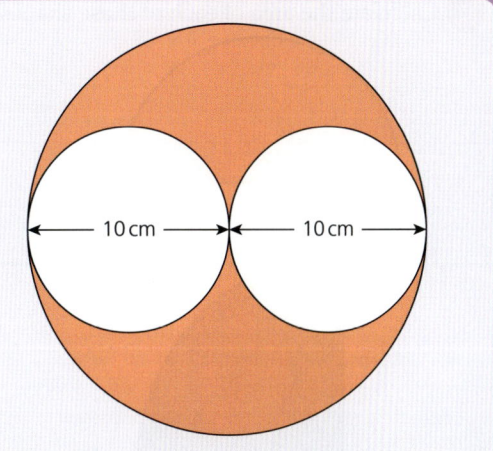

LET'S TALK

The diameter of each smaller circle is half that of the larger circle. What fraction of the larger circle is the area of each smaller circle?

Exercise 4.2

1. Calculate the area of each of these circles. Give your answers correct to one decimal place.

a

9 cm

b

0.9 cm

c

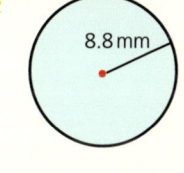

8.8 mm

d

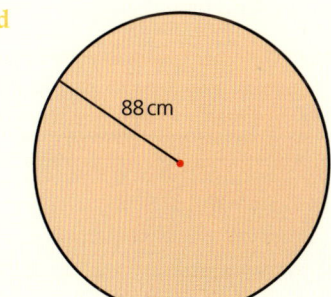

88 cm

e

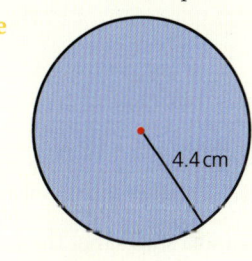

4.4 cm

f

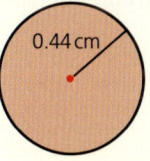

0.44 cm

2 Calculate the area of each of these shapes. Give your answers to a suitable degree of accuracy.

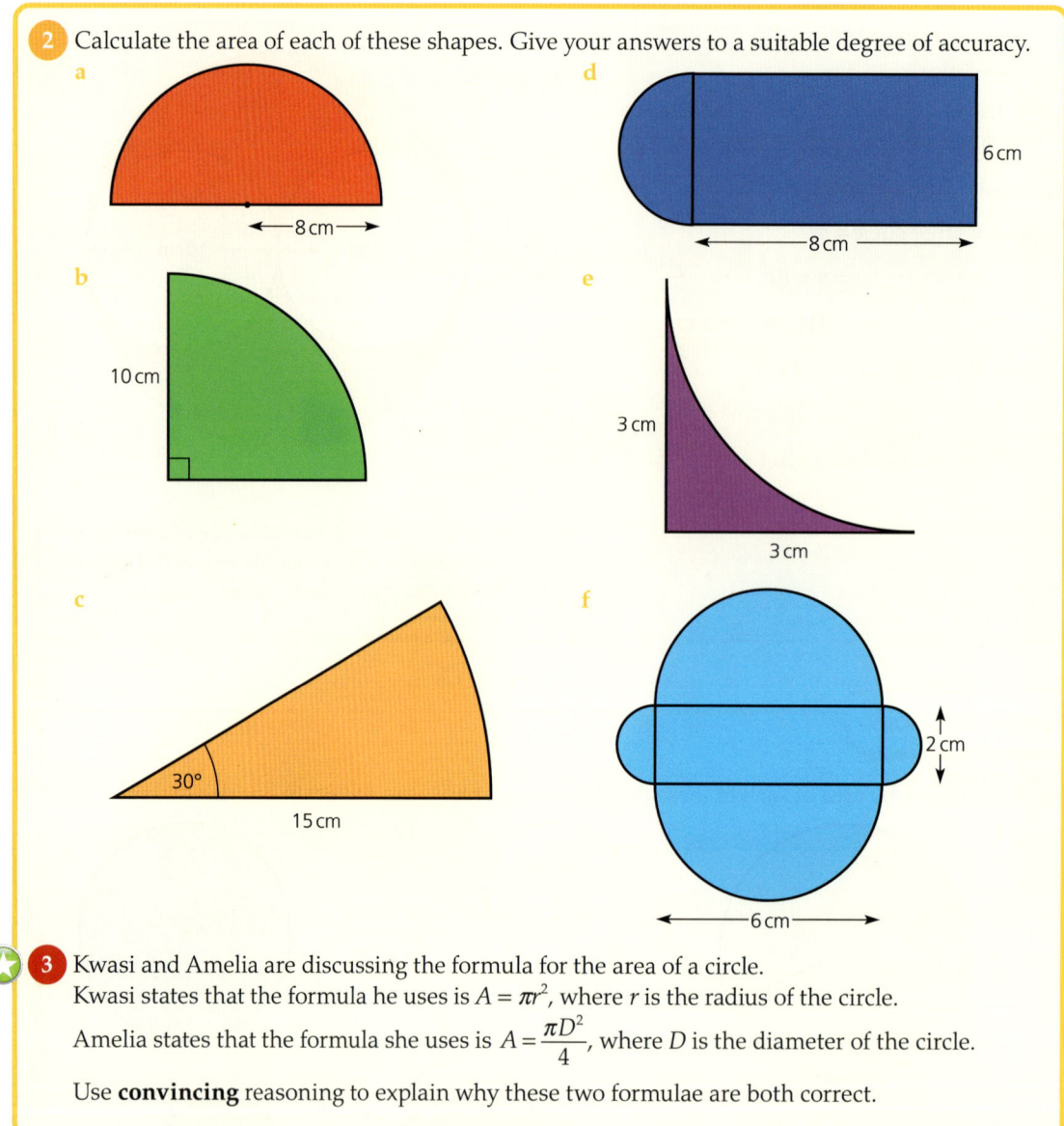

a

b 10 cm

c 30° 15 cm

d 6 cm 8 cm

e 3 cm 3 cm

f 2 cm 6 cm

3 Kwasi and Amelia are discussing the formula for the area of a circle.
Kwasi states that the formula he uses is $A = \pi r^2$, where r is the radius of the circle.

Amelia states that the formula she uses is $A = \dfrac{\pi D^2}{4}$, where D is the diameter of the circle.

Use **convincing** reasoning to explain why these two formulae are both correct.

Exercise 4.3

1 This diagram shows a circular disc inside a rectangular frame.
The width of the rectangle allows the disc just to fit inside the frame.

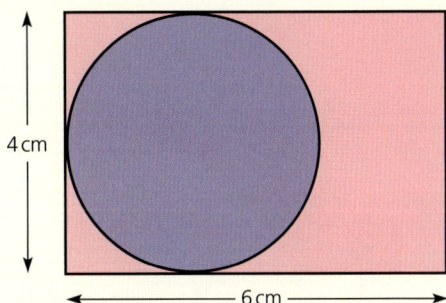

4 cm

6 cm

   a Calculate the area of the rectangle not covered by the disc.

   b What must the length of the rectangle be so that the area covered by the disc is the same as the area not covered by the disc?

2 Four semicircular pieces of chocolate are arranged in a box as shown in the diagram.

2 cm

Calculate:

   a the area occupied by the chocolate

   b the percentage of the base of the box not covered by the chocolate.

3 The diagram shows a circular disc inside a square frame.
The disc just fits inside the square.

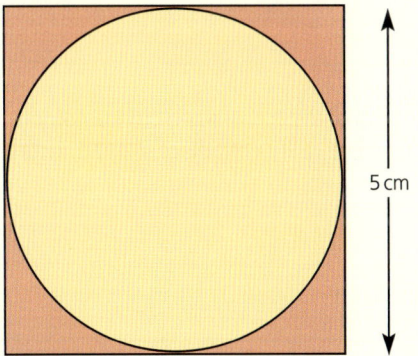

5 cm

   a Calculate the area of the square not covered by the disc.

b The square frame is kept the same size, but it is filled with four discs as shown.
Calculate the area of the square not covered by the discs.

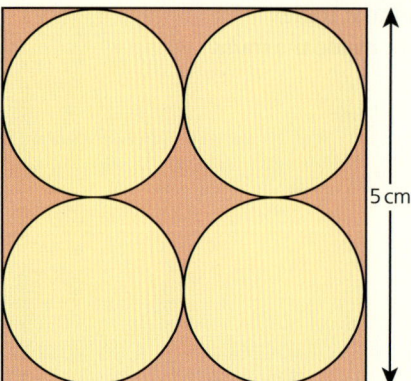

5 cm

c If the square frame is covered in 25 discs, explain how the area of the square not covered by the discs compares with your answers to parts (a) and (b).
Justify your answer.

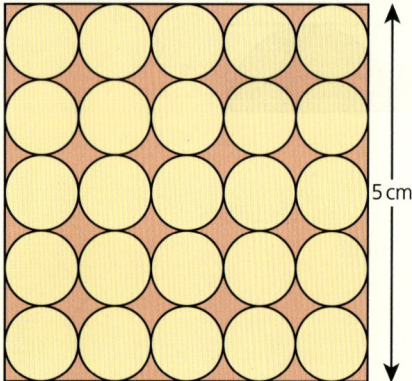

5 cm

4 This diagram shows a spiral made from five quadrants of different-sized circles, joined together. The radius of the smallest quadrant is 1 cm. The radius of adjacent quadrants doubles in size each time.

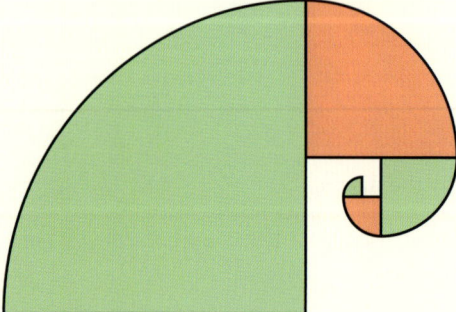

    a Calculate the area of the spiral. Give your answer correct to the nearest whole number of square centimetres.

    b Calculate the total perimeter of the spiral. Give your answer to the nearest whole number of centimetres.

5 A circular track has a width of 10 m. If the outer radius of the track is 100 m, calculate the area of the track. Give your answer to the nearest whole number of square metres.

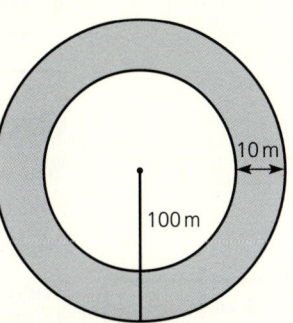

6 This diagram shows a target made from four concentric circles (circles with the same centre). The radius of the inner circle is 2 cm. The radius of the other circles increases by 2 cm each time.
Calculate the area of each of the rings numbered 1, 2 and 3. Give your answers to a suitable degree of accuracy.

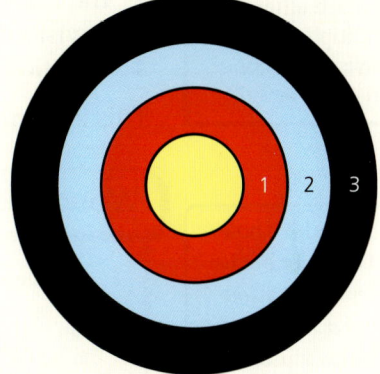

7 Pythagoras' thoerem states that for a right-angled triangle the square on the hypotenuse is equal to the sum of the squares of the other two sides, i.e. $a^2 = b^2 + c^2$ as shown by the diagram.
Does Pythagoras' theorem work when the areas of semicircles are used?

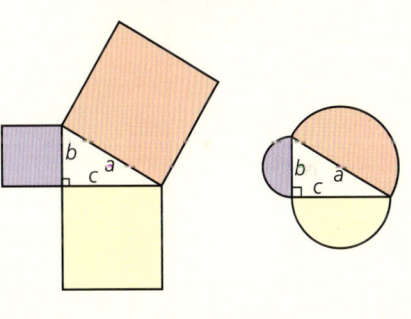

> Use your **specialising** skills by testing some numerical examples to see if Pythagoras' theorem still works.

Now you have completed Unit 4, you may like to try the Unit 4 online knowledge test if you are using the Boost eBook.

Order of operations with algebra

● Understand that the laws of arithmetic and order of operations apply to algebraic terms and expressions (four operations and integer powers).

Order of operations

You will already be aware from Stages 7 and 8 that when calculations are carried out using either numbers or algebra there is an order in which the calculations must be carried out.

LET'S TALK

Why, in respect to each other, can multiplication or division be done in either order?

The order is as follows:

1 **Brackets**
2 **Indices**
3 **Division/Multiplication**
4 **Addition/Subtraction.**

This is commonly written as **BIDMAS**.

Note here how the order of the multiplication and division could be swapped and the answer would remain unchanged.

LET'S TALK

When can the order be changed without affecting the final answer?

> ### Worked examples
>
> 1 Evaluate the following calculation: $\dfrac{4(3+7)^2}{8}$
>
> ● $\dfrac{4(10)^2}{8}$ *1st Brackets*
>
> ● $\dfrac{4\times100}{8}$ *2nd Indices*
>
> ● $\dfrac{400}{8}$ *3rd Multiplication*
>
> ● 50   *4th Division*
>
> 2 Evaluate the following when $x = 6$: $2(x-3)-\dfrac{x^2}{9}+2$
>
> ● $2(6-3)-\dfrac{6^2}{9}+2$ *1st Substitution*
>
> ● $2\times3-\dfrac{6^2}{9}+2$ *2nd Brackets*
>
> ● $2\times3-\dfrac{36}{9}+2$ *3rd Indices*
>
> ● $6-4+2$ *4th Multiplication and Division*
>
> ● 4   *5th Addition and Subtraction*

Exercise 5.1

1 Ollie and Gabriel are discussing the order in which they should carry out the following calculation: $4^2 + \left(\dfrac{15-7}{4}\right)^2 \times 3$

Ollie says that it's worked out in the following steps:

- $4^2 + \left(\dfrac{8}{4}\right)^2 \times 3$
- $4^2 + 2^2 \times 3$
- $16 + 4 \times 3$
- 20×3
- 60

Gabriel says that Ollie has made a mistake.

a Is there a mistake? Justify your answer.

b If there is a mistake, calculate the correct answer. Justify your answer.

2 Evaluate the following.

a $\dfrac{6^2}{3} - 4 \times 2$

c $\left(\dfrac{6}{3} - 4\right)^2 \times 2$

b $\left(\dfrac{6}{3}\right)^2 - 4 \times 2$

d $\left(\dfrac{6}{3} - 4 \times 2\right)^2$

3 Evaluate the following.

a $\dfrac{12-3}{3} + 5^2 \times 3$

d $\left(\dfrac{12-3}{3} + 5\right)^2 \times 3$

b $\dfrac{12}{3} - 3 + 5^2 \times 3$

e $\left(\dfrac{12-3}{3} - 5^2\right) \times 3$

c $12 - \dfrac{3}{3} + 5^2 \times 3$

4 a When $x = 6$, which of the following calculations produces the bigger answer?

i) $\dfrac{x(x+4)^2}{5} - \dfrac{16-x}{2}$

ii) $(3x+2)(2x-2) - \left(\dfrac{5x}{3}\right)^2$

b When $y = 4$, which of the following calculations produces the smaller answer?

i) $3y^2 + \dfrac{(2y-6)^2}{y-3} \times \dfrac{5y}{2^2}$

ii) $-3y + \dfrac{5y^2}{16} \times (3y+4)$

 Now you have completed Unit 5, you may like to try the Unit 5 online knowledge test if you are using the Boost eBook.

Large and small units

- Know and recognise very small or very large units of length, capacity and mass.

Large and small distances

Although there are lots of different units for length, the S.I. unit for length is the metre.

LET'S TALK
What other units of length do you know?

> **KEY INFORMATION**
> S.I. is short for Système Internationale and states the seven base units from which all other units are derived.

You will be familiar with the metre as a unit of measurement. Similarly, for longer distances, you will be used to using the kilometre. For smaller distances you will have used the millimetre.

However, these distances have limitations when distances are either very small or very large. In Unit 1 of this book you looked at the use of standard form to describe large and small numbers. But there are other units of measurement that can be used.

The distance from Earth to the Sun is approximately 149 million km, which can also be written as 149 000 000 km or 1.49×10^8 km.

We can just about visualise this distance in our heads.

LET'S TALK
Discuss what is meant by a trillion.

Apart from the Sun, the star nearest to Earth is called Proxima Centauri. Its distance from Earth is approximately 40 208 000 000 000 km.

This can also be written as 40.2 trillion km or 4.02×10^{13} km.

However, these numbers are more difficult to visualise.

A different unit is used when very large distances are being described, which is the **light year**.

KEY INFORMATION
Metres per second is usually shortened to either m/s or m s^{-1}.

Light can travel 299 792 458 metres in 1 second. To two significant figures and in standard form, this is 3.0×10^8 metres per second.

A light year is, as the name implies, how far light travels in 1 year and this is approximately equivalent to 9.46×10^{15} m.

> **KEY INFORMATION**
> The more accurate distance that light travels in 1 year is 9 460 730 472 580 800 m.

LET'S TALK
Describe the distance of 1 light year using the word 'trillion'.

The approximate distance from Earth to Proxima Centauri is therefore more commonly given as 4.25 light years.

> ## Worked example
>
> The distance of the star Alpha Centauri A from Earth is 4.13×10^{13} km.
>
> Taking 1 light year as 9.46×10^{15} m, how many light years is Alpha Centauri A from Earth?
>
> Changing km to m:
>
> 4.13×10^{13} km $= 4.13 \times 10^{16}$ m
>
> $4.13 \times 10^{16} \div 9.46 \times 10^{15} = 4.37$
>
> > Care must be taken as one of the distances is given in km, whilst the other is in m. For the calculation, both must be given in the same units.
>
> So, Alpha Centauri A is 4.37 light years from Earth.

This means that light would take 4.37 years to travel from Alpha Centauri A to Earth.

A light year is used for measuring very large distances.

For very small distances, though, the metre is split into smaller units. These are as follows:

- 1 **micrometre** = 1 millionth of a metre. Its symbol is μm.
  Therefore, 1 μm $= 1 \times 10^{-6}$ m.
- 1 **nanometre** = 1 billionth of a metre. Its symbol is nm.
  Therefore 1 nm $= 1 \times 10^{-9}$ m

LET'S TALK
Why is looking into space looking back in time?

KEY INFORMATION
A billion is 1000 million.

> ## Worked example
>
> The average human red blood cell has a diameter of approximately 8 μm.
>
> How many would need to be placed end to end in order to form a length of 1 cm?
>
> Both lengths need to be given in the same units, e.g. metres.
>
> 8 μm $= 8 \times 10^{-6}$ m
>
> 1 cm $= 0.01$ m
>
> $0.01 \div (8 \times 10^{-6}) = 1250$
>
> Therefore approximately 1250 red blood cells are needed to form a length of 1 cm.
>
>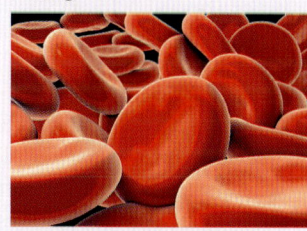

Exercise 6.1

1 Which of the following lengths are bigger than 1 cm?

    11 mm   0.02 m 8×10^{-3} m   2000 μm

2 Arrange the following lengths in order of size, starting with the smallest.

    420 mm   4 m   47 cm $\frac{1}{10}$th of a light year

    500 000 μn   400 000 nm

3 **a** Decide whether the following statement is true or false:

       A light year is a measure of time.

    b Justify your answer.

4 Taking the speed of light as 3.0×10^{8} m/s, convert the following into km.

    a 1.5 light years   **b** 8.245 light years

5 The distance from Earth to the Sun is approximately 149 million km. Taking the speed of light as 3.0×10^{8} m/s, calculate how long it takes sunlight to reach Earth.

    Give your answer in an appropriate unit.

6 **a** Write the following lengths in μm.

       i) 0.000 05 m   **ii)** 7.0×10^{-8} m   **iii)** 0.12 mm

    b Write the following lengths in nm.

       i) 5 μm   **ii)** 2.0×10^{-12} m

7 Taking the speed of light as 3.0×10^{8} m/s, work out which distance is bigger in each of the following pairs.

    a 2 000 000 000 000 km or 0.6 of a light year

    b 10 000 km or 1 billionth of a light year

8 Rearrange the following table into ascending order of length.

Object	Length
Amoeba	500 μm
Measles virus	220 nm
Grain of rice	6 mm
Human egg cell	0.000 13 m
Hepatitis virus	4.5×10^{-8} m
Sesame seed	3×10^{-3} m

9 Sirius is the brightest star in the night sky. It is 8.66 light years away. Taking the speed of light as 3.0×10^{8} m/s and the distance of the Sun from Earth as 149 million km, calculate how many times further Sirius is from Earth compared with the Sun.

    Give your answer to two significant figures.

10 The Large Hadron Collider at CERN in Geneva is circular, with a diameter of 8.5 km. It has managed to accelerate particles up to a speed of 299.8 million m/s. What is the maximum number of laps of the Collider that a particle can do in 1 second?

    Give your answer to two significant figures.

LET'S TALK

Does a light year take leap years into account?

KEY INFORMATION

A billion is 1000 million.

The Large Hadron Collider is the world's largest and most powerful particle accelerator. You may wish to investigate its uses and achievements further.

KEY INFORMATION

In ascending order means starting with the smallest.

Units for mass and capacity

Just as with distances, there are also other units used for large and small masses and also large and small **capacities**.

> **KEY INFORMATION**
> The word 'capacity' is a measure of volume dealing with liquids.

Mass

The S.I. unit for mass is the kilogram (kg), but other units of mass are related to it. These are:

- the **tonne** (t) – 1000 kilograms
- the gram (g) – 1 thousandth of a kilogram
- the **milligram** (mg) – 1 thousandth of a gram
- the **microgram** (μg) – 1 millionth of a gram.

Each of these can be written in kilograms using standard form:

- 1 tonne $= 1 \times 10^3$ kg
- 1 gram $= 1 \times 10^{-3}$ kg
- 1 milligram $= 1 \times 10^{-3}$ g and therefore 1×10^{-6} kg
- 1 microgram $= 1 \times 10^{-6}$ g and therefore 1×10^{-9} kg.

> ## Worked example
>
> A grain of sugar has an approximate mass of 0.000 625 g.
>
> **a** Write this mass in kg using standard form.
>
> As a gram is $\dfrac{1}{1000}$ of a kilogram, the mass must be divided by 1000:
>
> $0.000\,625 \div 1000 = 0.000\,000\,625$ kg
>
> In standard form, $0.000\,000\,625 = 6.25 \times 10^{-7}$ kg.
>
> **b** Write the mass in micrograms.
>
> As a gram is 1 000 000 times bigger than a microgram, the mass must be multiplied by 1 000 000:
>
> $0.000\,625 \times 1\,000\,000 = 625$
>
> Therefore, the mass of a grain of sugar $= 625$ μg.
>
> **c** Approximately how many grains of sugar are in a 1 kg bag?
>
> Using the same units:
>
> $1 \div (6.25 \times 10^{-7}) = 1\,600\,000$
>
> There are approximately 1 600 000 grains of sugar in a 1 kg bag.

Capacity

> **KEY INFORMATION**
>
> The prefix 'milli' means one thousandth of the unit that follows it, i.e. one millilitre is one thousandth of a litre.
>
> The prefix 'micro' means one millionth of the unit that follows it, i.e. one microlitre is one millionth of a litre.

Capacity deals with a volume of liquid.

The S.I. unit for capacity is the litre. As with other units, smaller quantities can be described using the millilitre (ml) and the **microlitre** (μl).

Therefore, as before:
- 1 millilitre=1 thousandth of a litre
- 1 microlitre=1 millionth of a litre.

In standard form these can be written as:
- 1 millilitre=1×10^{-3} litre
- 1 microlitre=1×10^{-6} litre.

Dividing by $1\,000\,000$ is the same as multiplying by 1×10^{-6}.

> Larger volumes are usually given in terms of m^3. However, this will not be covered in this book.

Worked example

A fine pipette can dispense 1.5 μl of liquid each time.

It is used to add liquid to 5000 test tubes.

How much liquid is added in total? Give your answer in:

a   litres

b   millilitres.

Total volume=$5000 \times 1.5 = 7500\,μl$

a   To convert μl to litres divide by $1\,000\,000$:

   Therefore $7500 \div 1\,000\,000 = 7.5 \times 10^{-3}$ litres.

b   To convert μl to millilitres divide by 1000:

   Therefore $7500 \div 1\,000 = 7.5\,ml$.

Exercise 6.2

1 For each of the following quantities, decide which symbol > , = or < must be placed between them to make the statement correct.

a 100 g   1 kg   **d** 1000 µg   1 mg

b 100 kg   1 t   **e** 80 ml   1 l

c 600 mg $\frac{1}{2}$ g   **f** 1000 µl $\frac{1}{100}$ l

2 The mass of a mosquito is given as 2.5×10^{-6} kg and the mass of a flea as 1.2×10^{-5} kg. Which is heavier? Justify your answer.

3 **a** How many grams are there in 2.5 t?
     b How many milligrams are there in 0.75 kg?
     c How many micrograms are there in 3.5 mg?
     d How many milligrams are there in 420 µg?
     e How many grams are there in 85 000 µg?

4 A housefly has a mass of 2.5×10^{-5} kg. Express this mass in:

a grams   **b** milligrams   **c** micrograms.

5 A teaspoon holds 5 ml of liquid.
How many teaspoons will be needed to fill a container of 420 litres?
Give your answer in standard form.

6 The average human cell has a mass of 1.0×10^{-12} kg.

a Express the mass in micrograms.
b If an adult has a mass of 81.5 kg, approximately how many cells will he have?
     Give your answer in standard form to 3 s.f.

7 1 cm³ of water has a mass of 1 gram.
1 litre of water has a mass of 1 kg.

a How many cm³ of water are equivalent to 1 litre?
b Calculate how many litres of water each of the following containers can hold.

    i) 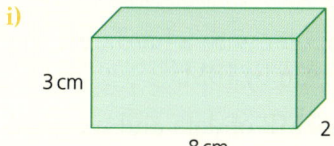   **ii)**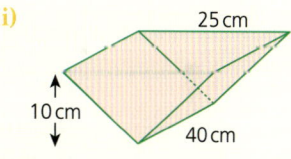

c Calculate the mass of water in each container.

8 A laboratory assistant can use two types of micropipette for an experiment. One type has a capacity of 4 µl, the other 15 µl.

a Which type of pipette should be used if exactly 0.14 ml is needed?
b **i)** Which type should be used if exactly 0.12 ml is needed?
     ii) Justify your choice in part (i).

LET'S TALK

Will 1 cm³ of all liquids and solids have the same mass?

How does this question relate to the Greek mathematician Archimedes?

9 The mass of $1\,\text{m}^3$ of water is given as 1 tonne.
The mass of 1 litre of water is $1\,\text{kg}$.
A large water container in the shape of a triangular prism is shown (not to scale).

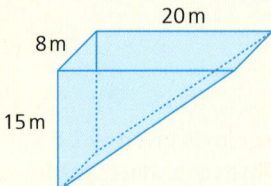

20 m

8 m

15 m

  a Calculate the mass of water needed to fill the container.
  b If water is pumped in at a rate of 220 litres per minute, calculate how long it will take to fill.
     Give your answer using suitable units and justify your choice of units.

Digital storage

The past 30 years have seen an explosion in the use of the computer in every aspect of our lives. From the storage in a phone to the storage capabilities of large digital companies, the ability to store huge amounts of data has become a central part of modern life.

The unit of digital information is known as the **byte** (B).

Multiples of this unit are as follows:
- the **megabyte** (MB) = 1 million bytes, i.e. 1×10^6 B
- the **gigabyte** (GB) = 1 billion bytes, i.e. 1×10^9 B
- the **terabyte** (TB) = 1 trillion bytes, i.e. 1×10^{12} B.

The first computer with a hard disk was produced in 1956 by IBM.

It could store $5\,\text{MB}$ of data and weighed over a tonne.

Today, home computers with a storage capability of $1\,\text{TB}$ are not uncommon.

▲ The IBM 350 being transported by plane.

LET'S TALK

If you have a smartphone, compare its storage capacity with that of the IBM 350.

How much more data can your phone hold?

Worked example

A memory card for a camera has a storage capability of 32 GB.

The digital camera takes photographs with an average file size of 12 MB.

How many photographs can be stored on the memory card?

Converting 32 GB into MB so that we are working with the same units:

1 GB = 1000 MB, therefore 32 GB = 32 000 MB

32 000 ÷ 12 = 2667 (to 4 s.f.)

Therefore, the memory card can hold approximately 2667 photographs.

Exercise 6.3

1   Arrange the following measures of digital storage in order of size, starting with the largest.

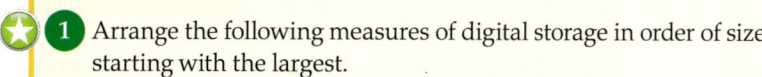

   10 MB $\frac{1}{10}$ GB   1 millionth of a TB   1.2 million B

2   Anuska is trying to convert 4500 MB into GB.
    Her method is as follows:
    100 MB = 1 GB
    Therefore 4500 MB = 45 GB (because 4500 ÷ 100 = 45)
    Critique what she has done wrong and work out the correct conversion.

3   Convert the following storage quantities.

   a   10 000 000 B to MB   d $\frac{1}{5}$ TB to MB
   b   500 MB to GB
   c   50 GB to TB   e   0.45 GB to MB

4   A dual-layer Blu-ray disc can store 50 GB of data.
    How many Blu-ray discs can be stored on a computer with a 4 TB capacity?

5   The book *War and Peace* by Leo Tolstoy is one of the longest novels written. When converted to digital format it occupies approximately 1.7 MB of storage.
    How many copies of *War and Peace* can be stored on a 1 GB hard drive?

6   A phone can hold 16 GB of data.
    An 8.3 GB video and 620 photos each of 6 MB are stored on the phone.
    The apps occupy a further 2.4 GB of storage.
    a   How many GB of storage are still left on the phone?
    b   Convert your answer to part (a) into MB.

7 A digital camera has a memory card with a storage capacity of 8 GB. The camera can take high resolution photographs with an average file size of 12 MB, medium resolution photographs with an average file size of 5 MB or low resolution photographs with an average file size of 1.5 MB.

The memory card already has the following numbers of photographs on it:

- 210 high resolution
- 424 medium resolution
- 1210 low resolution.

The photographer is not sure whether to take the rest of his photographs on a medium or a low resolution setting.

How many fewer photographs can he take if he chooses the medium resolution setting rather than the low resolution setting?

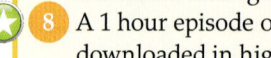 **8** A 1 hour episode of a series uses up 440 MB of storage when downloaded in high definition.

Freya and Matthew want to download all the episodes from all the series of a popular TV programme.

There are six episodes per series, and eight series have been produced. Their computer has 25 GB of available storage.

a Can they download all of the programmes? Justify your answer.

b They start to download the programmes at 8 p.m.
If their download speed is 60 MB per second, at what time will all the programmes have been downloaded?

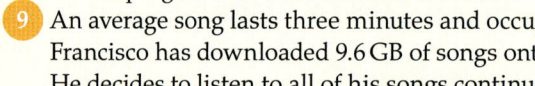 **9** An average song lasts three minutes and occupies 4 MB of storage. Francisco has downloaded 9.6 GB of songs onto his phone.

He decides to listen to all of his songs continuously in one go.

He starts playing them at 9.00 a.m. on 1st January.

At what time and date will he have finished listening to all of his songs?

 Now you have completed Unit 6, you may like to try the Unit 6 online knowledge test if you are using the Boost eBook.

7 Record, organise and represent data

- Record, organise and represent categorical, discrete and continuous data.
- Choose and explain which representation to use in a given situation.

Interpreting and discussing results

You are already familiar with many ways of recording and displaying data, including tables, frequency charts, pie charts, line graphs and stem-and-leaf diagrams. In addition to recording and organising data, it is also important to be able to interpret results that have been presented in these ways.

Recap

Scatter graphs and correlation

Scatter graphs are particularly useful if we wish to see if there is a correlation (relationship) between two sets of discrete data.
The two sets of data are collected. Each pair of values then form the coordinates of a point, which is plotted on a graph.

There are several types of correlation, depending on the arrangement of the points plotted on the scatter graph.

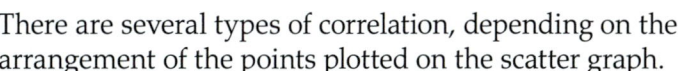

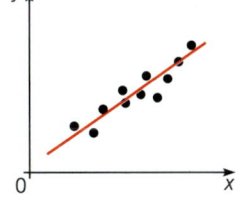

A strong positive correlation between the values of x and y. The points lie very close to the **line of best fit**.

As x increases, so does y.

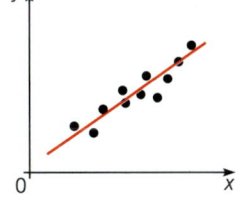

A weak positive correlation. Although there is direction to the way the points are lying, they are not tightly packed around the line of best fit.

As x increases, y tends to increase too.

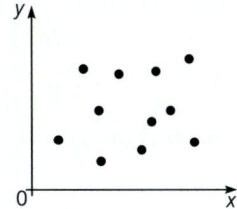

No correlation. As there is no pattern in the way the points are lying, there is no correlation between the variables x and y. As a result there can be no line of best fit.

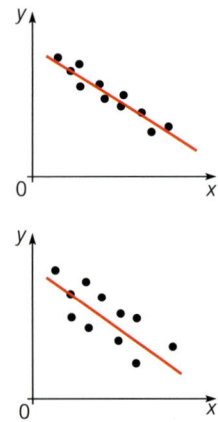

A strong negative correlation. The points lie close to the line of best fit.

As x increases, y decreases.

A weak negative correlation. There is direction to the way the points are lying but they are not tightly packed around the line of best fit.

As x increases, y tends to decrease.

Worked example

A group of students believed that the time students took to get to school depended largely on how far away they lived. They collected data to test this theory, recording how far each student lived from school (in kilometres) and the time it took them to get to school (in minutes).

This is a graph of their results.

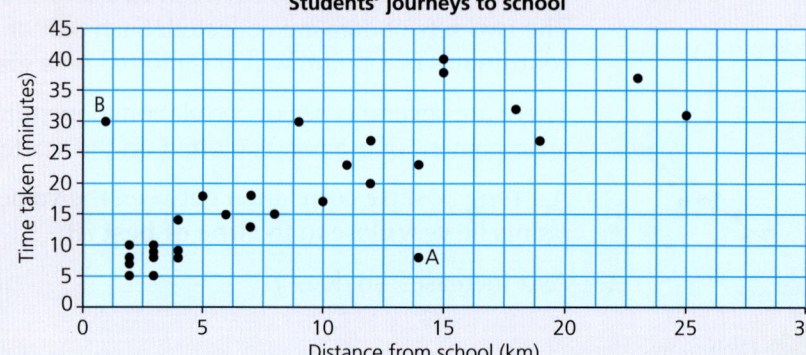

LET'S TALK

What other pairs of data would also be likely to produce a positive correlation?

Which would probably give a negative correlation?

a What type of graph has been plotted?

A scatter graph.

b Does the graph support the students' theory that the further away a student lives from school, the longer it takes to travel?
Justify your answer.

Yes, as there is a positive correlation with the points tending to go from bottom left up to top right.

c Explain why some students who live further away may get to school more quickly than other students who live nearer.

They may come by car, whilst a student who lives relatively close may choose to walk to school.

d The data points for two students, A and B, have been highlighted on the graph.

   i) What does the position of point A tell you about student A's distance from school and travel time?

     Student A lives fairly far away, but gets to school quickly.

   ii) What does the position of point B tell you about student B's distance from school and travel time?

     Student B lives close to school but takes a relatively long time to get to school.

e Draw a line of best fit on the graph and use it to estimate how long it would take a student who lives 15 km away to get to school.

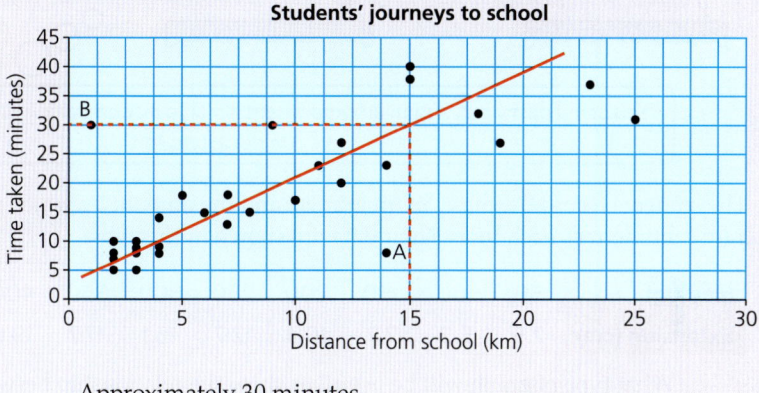

Approximately 30 minutes.

LET'S TALK

Why does a line of best fit not always pass through the origin? Give some examples of when this is the case.

Exercise 7.1

1 The following infographic incorporates two pie charts.

How often students cycle each week
Yr.7 compared with Yr.11

- Never
- Once
- Twice
- Three or more

Yr.7   Yr.11

Critique the infographic. Make sure you comment on
a two strengths   **b** two weaknesses.

2 In a science experiment, a spring is attached to a clamp stand as shown. Different masses are hung from the end of the spring.

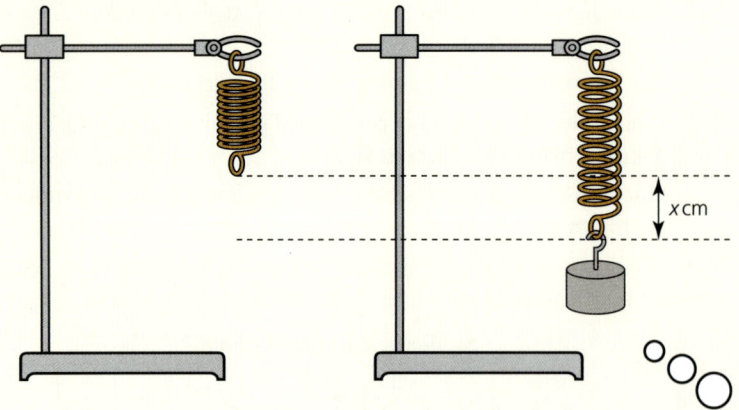

> You could carry out a similar experiment to this by comparing different thicknesses of elastic bands.

The students record the mass, m (in grams), and the amount by which the spring stretches, x (in centimetres), each time. Their data are shown in this table.

Mass (g)	50	100	150	200	250	300	350	400	450	500
Extension (cm)	2.5	5.2	7.2	10.4	12.7	15.1	17.7	19.6	22.0	24.4

a i) What type of graph will be useful to show the relationship between the two variables?
  ii) Plot a graph of extension against mass.
b Describe the correlation (if any) between the extension and the mass.
c Draw a line of best fit for the data.
d Use your graph to predict the extension of the spring if a mass of 375 g is hung from it.
e A student extends the line of best fit to predict what the extension will be if 4 kg is hung from the spring.
  Explain why this prediction is not likely to be accurate.

3 A newspaper report states that:
  'On average, women live longer than men.'
The United Nations keeps an up-to-date database of statistical information on its member countries. Some of the data, giving the life expectancies of men and women, are shown in this table.

	Life expectancy at birth (years, 1990–99)	
Country	Female	Male
Australia	81	76
Barbados	79	74
Brazil	71	63
Chad	49	46
China	72	68
Colombia	74	67
Congo	51	46
Cuba	78	74
Egypt	68	65
France	82	74
Germany	80	74
India	63	62

	Life expectancy at birth (years, 1990–99)	
Country	Female	Male
Iraq	64	61
Israel	80	76
Japan	83	77
Kenya	53	51
Mexico	76	70
Nepal	57	58
Portugal	79	72
Russian Federation	73	61
Saudi Arabia	73	70
UK	80	75
USA	80	73

One of the newspaper's readers wants to check the statement in the newspaper. They use the United Nations data to plot this scatter graph.

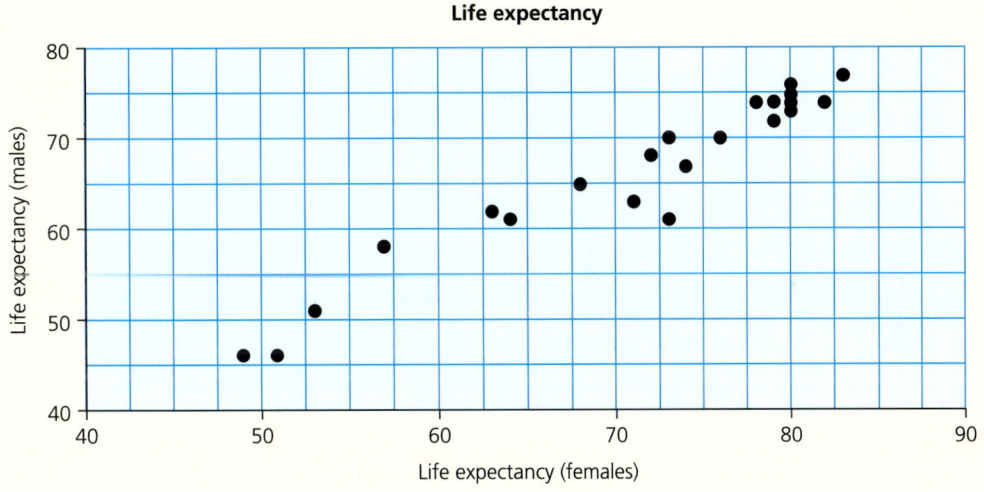

a Is there a correlation between the life expectancies of males and females? Justify your answer.

b Do the data and the graph support the statement made by the newspaper? Justify your answer.

4 The graph below shows the distribution of the masses of birds of two different species, P and Q. The mean mass of a bird from species P is M_1; the mean mass of a bird from species Q is M_2.

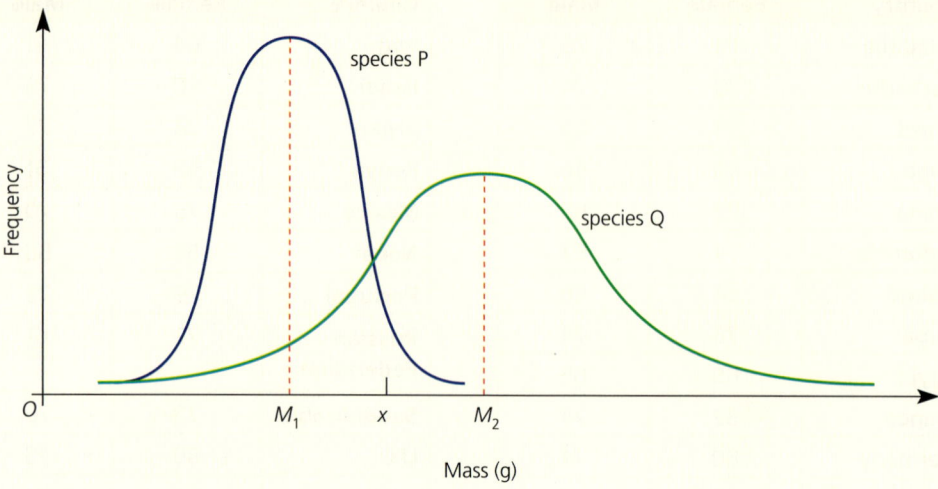

Mass (g)

a Which species has the larger mean mass?
b Which species has masses which are within a smaller range? Justify your answer.
c Another bird is caught and its mass is measured. Its mass is x g (shown on the graph). Is this bird more likely to be from species P or species Q? Justify your answer.

5 The manufacturers of two different brands of batteries, X and Y, collect data about how long their batteries last under 'average' usage. This graph shows their data.

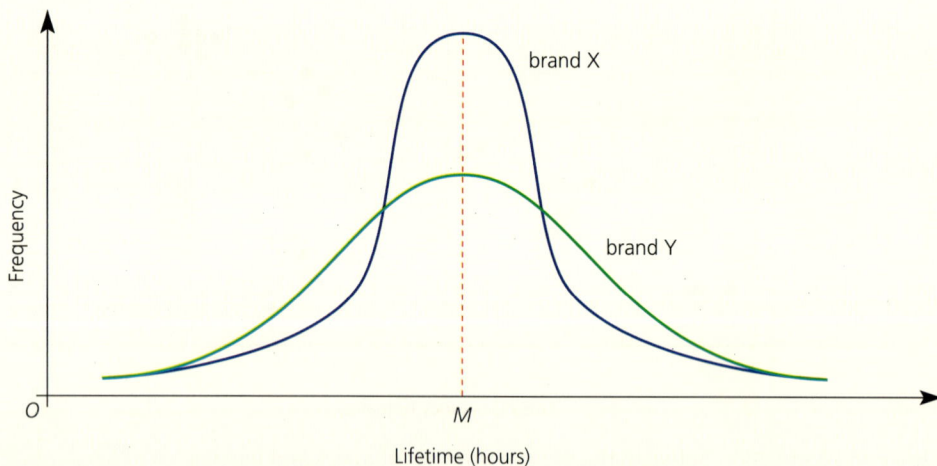

Lifetime (hours)

Both brands of battery have the same mean lifetime, M, as shown.
Which brand is more consistent? Justify your answer.

6 This graph shows the distributions of the ages of the members of two clubs, A and B.
One is an athletics club; the other is a golf club.

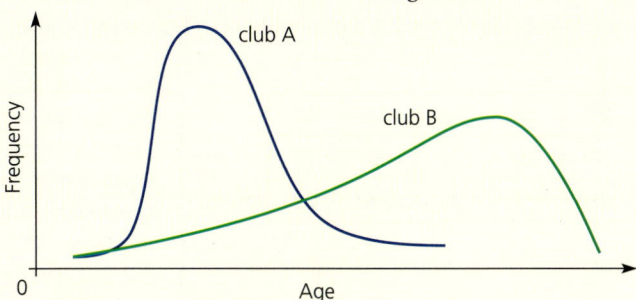

Which club is more likely to be the golf club? Justify your answer.

7 The heights of competitors at a sporting event are shown in the table.

Height (cm)	Frequency
$160 \leqslant h < 165$	6
$165 \leqslant h < 170$	6
$170 \leqslant h < 172$	6
$172 \leqslant h < 173$	6
$173 \leqslant h < 175$	6
$175 \leqslant h < 185$	6

a What type of data is height?
b Explain what is meant by the inequality $160 \leqslant h < 165$.
c One observer states that the data show that the heights of the competitors are evenly distributed. **Critique** their statement.

Stem-and-leaf diagrams

You will already be familiar with the fact that stem-and-leaf diagrams are used for discrete data (or continuous data that has been rounded) and are a special type of bar chart in which the 'bars' are made from the data itself. This has the advantage that the original data can still be seen in the diagram.

In Stage 8 you looked at how to present these data of the ages of people on a bus to the seaside using a stem-and-leaf diagram.

2	4	25	31	3	23	24	26	37	42
60	76	33	24	25	18	20	77	5	13
18	13	15	49	70	48	27	25	24	29

The diagram must have a key to explain what the stem means. If the data were 1.8, 2.7, 3.2 etc., the key would state that '2 | 7 means 2.7'.

Displaying the data on a stem-and-leaf diagram produced the following:

```
0 | 2   3   4   5
1 | 3   3   5   8   8
2 | 0   3   4   4   4   5   5   5   6   7   9
3 | 1   3   7
4 | 2   8   9
5 |
6 | 0
7 | 0   6   7
```

Key

3 | 1 means 31 years

Stem-and-leaf diagrams can also be used to compare two different sets of data by arranging them 'back to back'.

Worked example

Another bus picks up 30 people from a care home to take them to a local supermarket. Their ages are:

72	74	65	81	83	73	64	66	77	72
60	76	83	84	85	68	80	77	75	73
88	73	75	79	70	88	67	65	74	69

Display these data on a back-to-back stem-and-leaf diagram with the data for the seaside trip from the example at the top of this page.

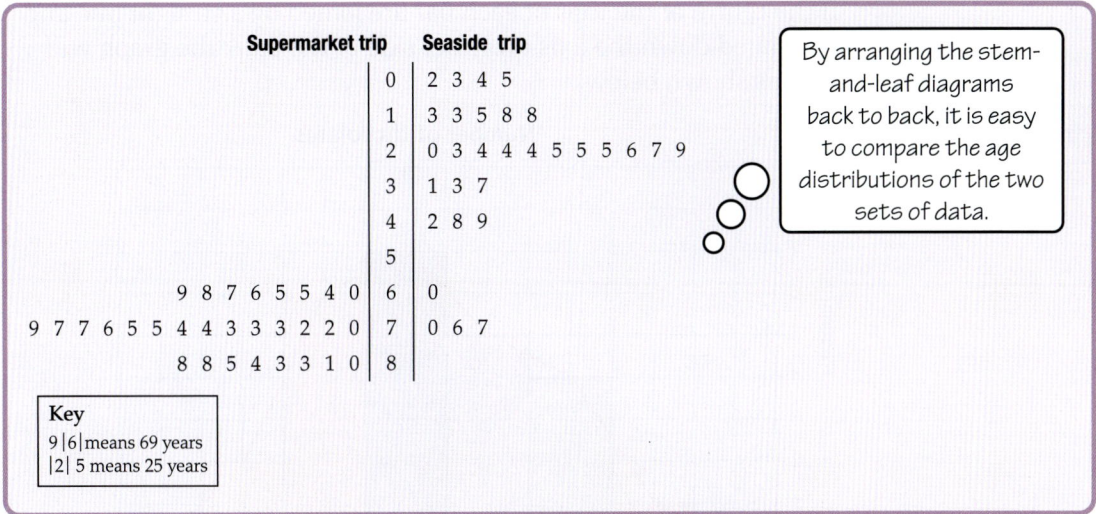

	Supermarket trip		Seaside trip
		0	2 3 4 5
		1	3 3 5 8 8
		2	0 3 4 4 4 5 5 5 6 7 9
		3	1 3 7
		4	2 8 9
		5	
	9 8 7 6 5 5 4 0	6	0
9 7 7 6 5 5 4 4 3 3 3 2 2 0	7	0 6 7
	8 8 5 4 3 3 1 0	8	

By arranging the stem-and-leaf diagrams back to back, it is easy to compare the age distributions of the two sets of data.

Key
9 | 6 | means 69 years
| 2 | 5 means 25 years

By arranging the stem-and-leaf diagrams back to back, it is easy to compare the age distributions of the two sets of data.

Frequency polygons

Another way of representing frequency diagrams is to use a **frequency polygon**.

The frequency diagram below shows the number of chocolates found to be in 25 tins of a particular brand.

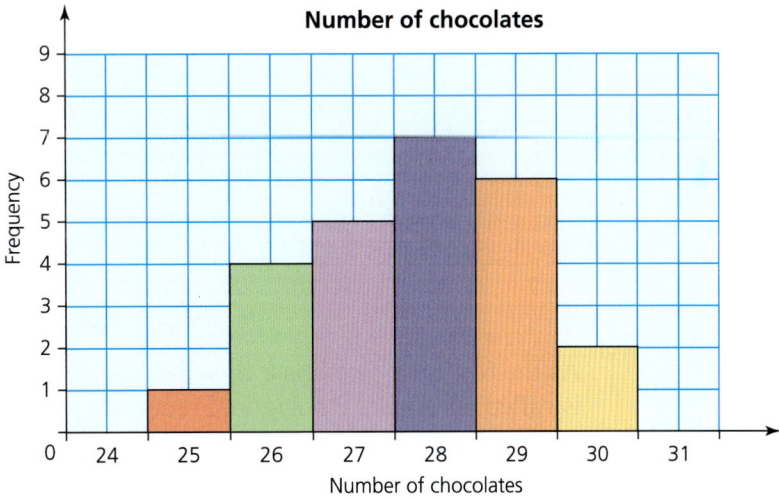

Number of chocolates

If the centre of the tops of each bar are joined by a straight line and continued down the axis on either side of the data, the graph will look as shown below.

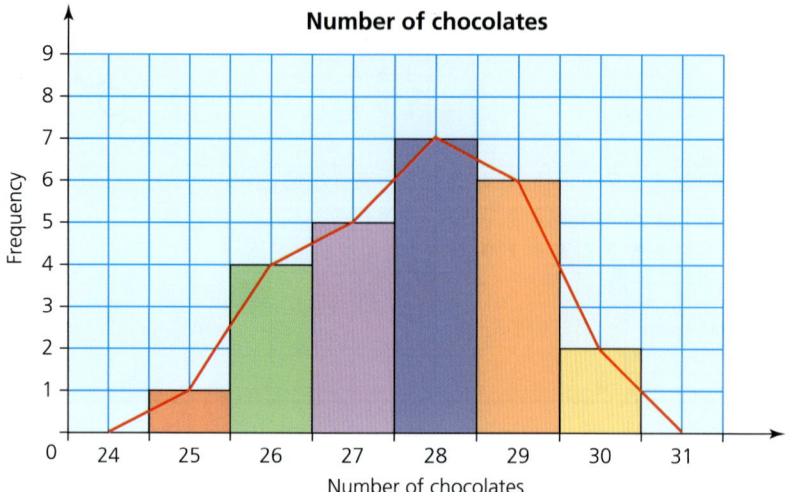

By removing the bars, the look of the graph is simplified, without losing any of its information, as shown.

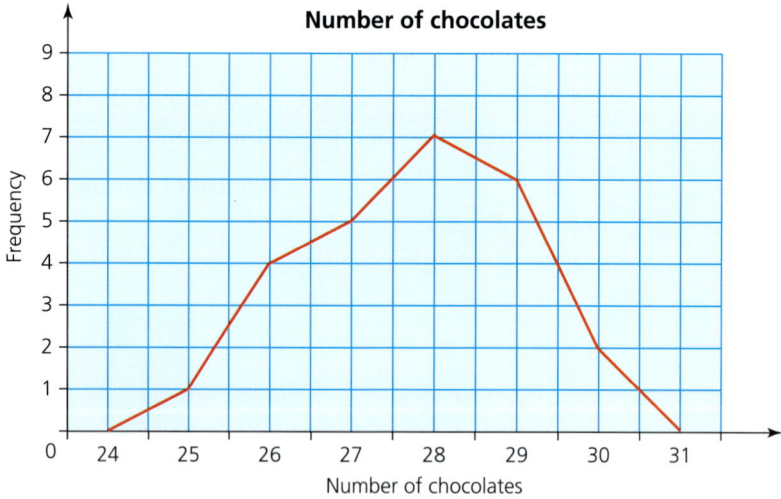

As the resulting graph is a closed shape, made up of straight lines, it is called a frequency polygon. Frequency polygons are useful as they are quicker to draw than frequency diagrams, but show the same information.

Exercise 7.2

 1 Twenty students sit two mathematics tests, A and B. Each test is marked out of 50. The results are shown below.

Test A									
8	10	12	16	20	21	21	23	24	26
29	34	34	35	36	36	38	38	41	42

Test B									
18	26	27	27	29	34	34	35	35	36
38	38	39	41	41	43	45	48	48	49

 a   Display the data on a back-to-back stem-and-leaf diagram.
 b   From the diagram, which test appears to have been the harder of the two? Justify your answer.

2 A basketball team plays 25 matches during one season. They keep a record of the number of points they score in each game and the number of points scored against them. The results are shown below.

Points for					Points against				
64	72	84	46	53	51	72	40	66	69
69	62	71	71	79	43	58	81	78	60
80	47	53	69	69	60	42	57	69	74
56	82	84	78	78	40	41	72	66	54
72	68	66	54	64	72	66	51	53	67

 a   Draw a back-to-back stem-and-leaf diagram showing the number of points for and against the team during the season.
 b   By looking at the shape of the diagram, is the team likely to have won more games than lost, or the other way around? Justify your answer.

3 The mass (m kg) of 50 newborn babies are given in the grouped frequency table below. Plot a frequency polygon for this data.

Mass (kg)	Frequency
$1.5 < m \leqslant 2.0$	0
$2.0 < m \leqslant 2.5$	1
$2.5 < m \leqslant 3.0$	4
$3.0 < m \leqslant 3.5$	12
$3.5 < m \leqslant 4.0$	18
$4.0 < m \leqslant 4.5$	10
$4.5 < m \leqslant 5.0$	3
$5.0 < m \leqslant 5.5$	2
$5.5 < m \leqslant 6.0$	0

4 Fifteen people take part in a fitness assessment. Their pulse rates are taken before and after exercise. The results are recorded in this back-to-back stem-and-leaf diagram.

```
        4   3   1 | 5 |
            9   6 | 5 | 8
    4   3   3   0 | 6 | 1   3
        8   8   7 | 6 | 8   8   9
                1 | 7 | 0   2   3   4
            9   9 | 7 | 7
                  | 8 | 1   3
                  | 8 | 9
                  | 9 | 3
```

Key
4 | 5 | means 54 beats per minute
| 6 | 8 means 68 beats per minute

a Calculate the mean, median, mode and range for the pulse rates for both sides of the diagram.

b Which side of the diagram is likely to show the readings taken after exercise? Justify your answer using your answers in part (a) to support your reason.

5 Students from two classes took the same test.
This back-to-back stem-and-leaf diagram shows their results.

```
Class X              Class Y

    8   8   6   6 | 1 | 1
        7   4   3   2 | 2 | 6   7   9
8   6   6   4   4 | 3 | 4   5   5   5   5   6   6   7   8   9   9   9   9
        9   9   7   7 | 4 | 6   8   8
        8   5   5   5 | 5 | 2
```

Key
6 | 1 | 1 means 16 marks in class X and 11 marks in class Y

Your teacher will be able to give you the test results from two different classes for you to analyse in a similar way.

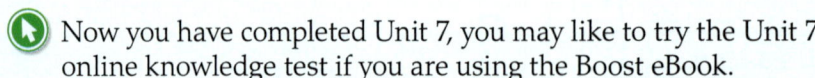

a i) Why is a back-to-back stem-and-leaf diagram a useful graph to use?
   ii) Compare the distribution of results from the two classes, commenting on any similarities and/or differences.

b One class of students is set by ability and the other is a mixed ability class.
From the results, deduce which class is likely to be set by ability. Justify your answer.

▶ Now you have completed Unit 7, you may like to try the Unit 7 online knowledge test if you are using the Boost eBook.

8 Surface area and volume of prisms

- Use knowledge of area and volume to derive the formula for the volume of prisms and cylinders. Use the formula to calculate the volume of prisms and cylinders.
- Use knowledge of area, and properties of cubes, cuboids, triangular prisms, pyramids and cylinders to calculate their surface area.

Volume of a prism

A prism is a three-dimensional shape which has the same **cross-sectional area** all through it, i.e. if you were to cut slices through the shape, the shape and area of each slice would be exactly the same.

Here are some examples of prisms:

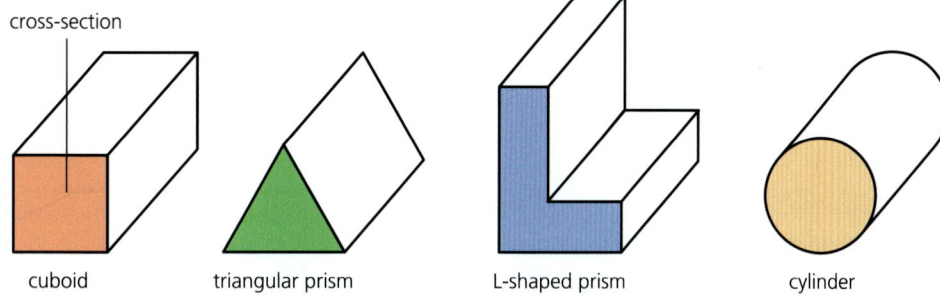

cuboid   triangular prism   L-shaped prism   cylinder

LET'S TALK

If these shapes were sliced from a different direction, would the cross-sectional area always be constant?

When each of these shapes is sliced parallel to the coloured face, the cross-section will always look the same. To calculate the volume of a prism:

Volume of a prism = area of cross-section × length

A cylinder is a type of prism as, if sliced parallel to the end face, the cross-section would be the same. The shape of the constant cross-section for a cylinder is a circle.

From the work covered in Unit 4, we know that the area of a circle is given by πr^2, where r is the radius of the circle. Therefore:

Volume of a cylinder = $\pi r^2 \times$ length

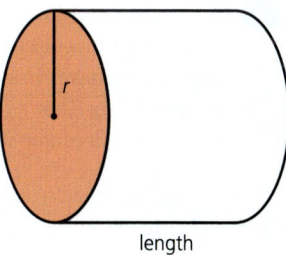

length

Worked examples

1 Calculate the volume of this cuboid.

Area of cross-section $= 6 \times 4 = 24\,cm^2$

Volume $= 24 \times 3 = 72\,cm^3$

In this case, because the prism is a cuboid, other cross-sections could have been used, for example:

Area of cross-section $= 3 \times 4 = 12\,cm^2$

Volume $= 12 \times 6 = 72\,cm^3$

The volume is the same as before.

2 Calculate the volume of this cylinder.

Give your answer correct to one decimal place.

Area of cross-section $= \pi \times 5^2 = 78.539\,816\ldots\,cm^2$

Volume $= 78.539\,816\ldots \times 10 = 785.4\,cm^3$ (to 1 d.p.)

3 Calculate the volume of this 'N'-shaped prism.

Area of cross-section $=$ area A $+$ area B $+$ area C

$= (2 \times 5) + (3 \times 3) + (2 \times 5)$

$= 10 + 9 + 10$

$= 29\,cm^2$

Volume $= 29 \times 8 = 232\,cm^2$

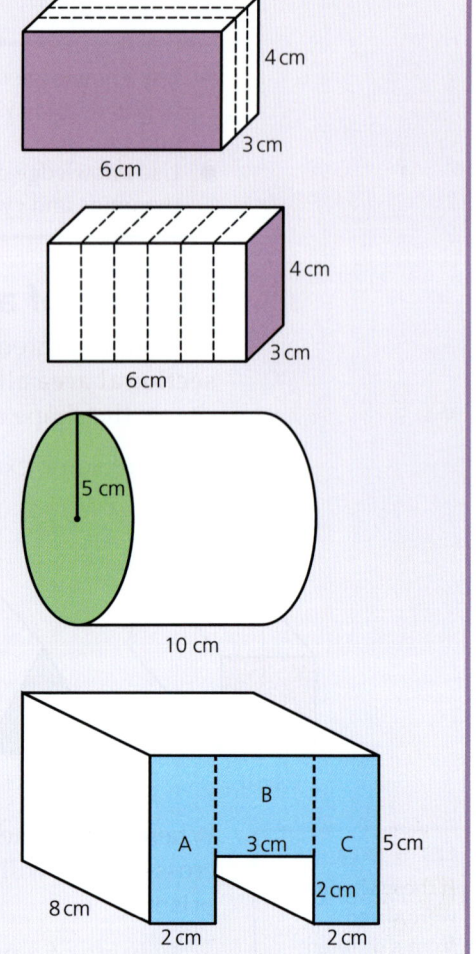

Exercise 8.1

1 Calculate the volume of each of these cylinders, where $r=$ radius of circle, $D=$ diameter and $L=$ length. Give your answer in the same units as those used for D or r in each case. (Use your calculator value for π.)

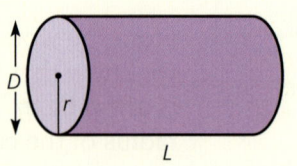

a $r=3\,cm$	$L=12\,cm$
b $r=6\,cm$	$L=25\,cm$
c $D=9\,cm$	$L=0.15\,m$
d $D=58\,mm$	$L=0.35\,m$
e $r=\pi\,cm$	$L=50\,mm$

2 Calculate the volume of each of the following prisms.

a

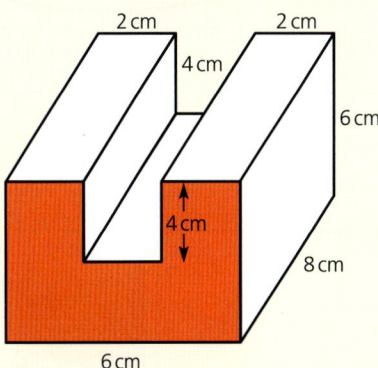

2 cm   2 cm

4 cm

6 cm

4 cm

8 cm

6 cm

b

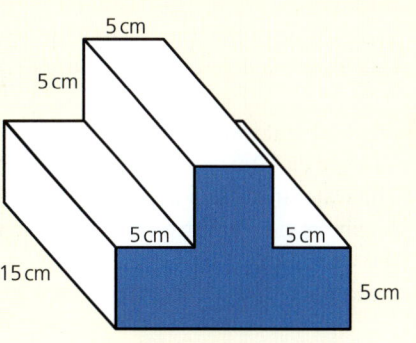

5 cm

5 cm

5 cm   5 cm

15 cm

5 cm

3 Which of the following prisms have a volume greater than 100 cm³?

a

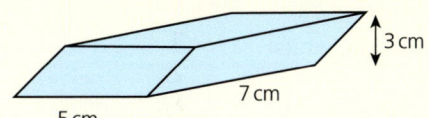

3 cm

7 cm

5 cm

b

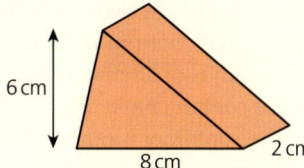

6 cm

8 cm

2 cm

c

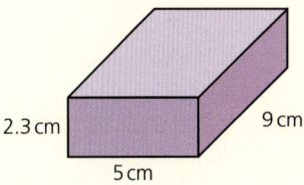

2.3 cm

9 cm

5 cm

4 Part of a steel pipe is shown. The inner radius is 60 mm, the outer radius is 65 mm and the pipe is 200 m long. Calculate the volume of steel used in making the pipe. Give your answer in cm³ and correct to three significant figures.

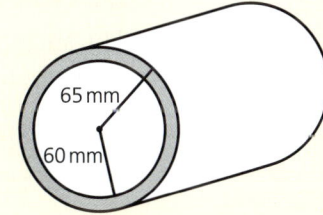

65 mm

60 mm

5 A cylinder has a volume of 1540 cm³.
If the radius of the circular cross-section is 7 cm, calculate the cylinder's length L.

$\left(\text{Take } \pi = \dfrac{22}{7}\right)$

LET'S TALK

$\dfrac{22}{7}$ is often used as an approximation for π.
How accurate is it?

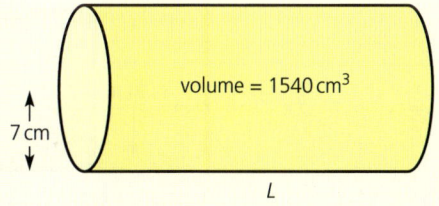

7 cm

volume = 1540 cm³

L

6 A disc is shown in the diagram. It has a thickness of 1 mm. The diameter of the disc is 12 cm. The volume of the disc is 11.13 cm³ (to two decimal places). Calculate the diameter of the inner circle.

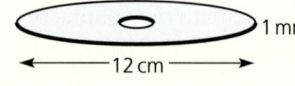

7 A cylindrical water tank has a radius of 40 cm and a height of 1.2 m.
The water in it has a depth of 60 cm.
A cube of side length 50 cm is placed at the bottom of the water tank.
How much does the depth of the water increase by?

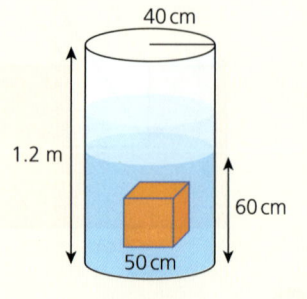

8 A water tank consists of two cylindrical sections. Both have a depth of 80 cm. The bottom section has a radius of 50 cm, whilst the top has a radius of 20 cm as shown.
A tap fills the container at a rate of 5 litres per minute.
Assuming the container is empty to start with, work out how long it will take to fill.
Give your answer to the nearest minute.

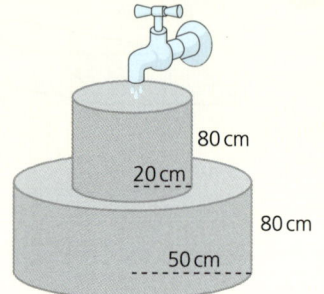

> **KEY INFORMATION**
> 1 litre volume is equivalent to 1000 cm³.

Surface area of a prism

The surface area of a prism (indeed of any three-dimensional shape) is the total area of its faces.

Worked examples

1 Calculate the surface area of this cylinder.

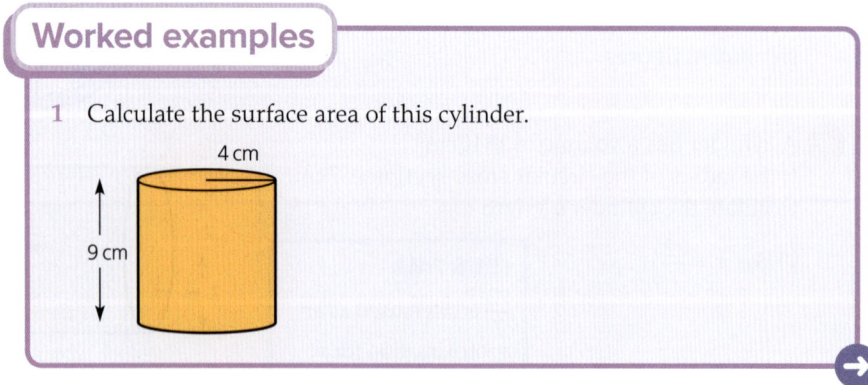

Draw (or imagine) the net.

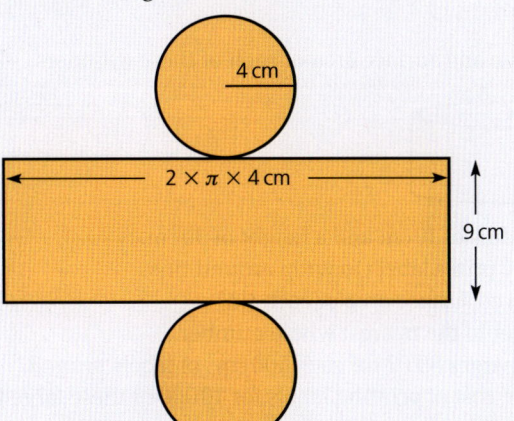

Write down your answers rounded to two decimal places at each stage but store the full answers in your calculator's memory. Use these for subsequent calculations to avoid rounding errors.

The surface area of a cylinder is made of two circles and a rectangle. The length of the rectangle is equivalent to the circumference of one of the circles.

Area of one circle = $\pi \times 4^2 = 50.27\,\text{cm}^2$

Area of both circles = $50.27\ldots \times 2 = 100.53\,\text{cm}^2$

Area of rectangle = $2 \times \pi \times 4 \times 9 = 226.19\,\text{cm}^2$

Total surface area = $226.19 + 100.53 = 326.73\,\text{cm}^2$ (to 2 d.p.)

2   Calculate the total surface area of this triangular prism.

The triangular prism has five faces, therefore the total surface area is the sum of the areas of the five faces.

Area of each triangular face = $\frac{1}{2} \times 12 \times 5 = 30\,\text{cm}^2$

Therefore, area of both triangular faces = $30 \times 2 = 60\,\text{cm}^2$

Base of the prism is a rectangle. Its area is $12 \times 15 = 180\,\text{cm}^2$

Back face of the prism is a rectangle. Its area is $15 \times 5 = 75\,\text{cm}^2$

The sloping face of the prism is also a rectangle. However, one of its dimensions is not given. The length of the sloping edge needs to be calculated. As it is the hypotenuse of the right-angled triangle it can be calculated using Pythagoras' theorem:

$H^2 = 5^2 + 12^2$

$H^2 = 169$

$H = \sqrt{169} = 13\,\text{cm}$

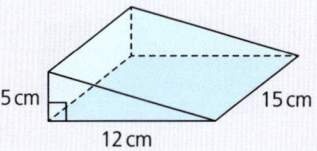

Therefore, the area of the rectangular sloping face is $13 \times 15 = 195\,\text{cm}^2$.

Total surface area = $60 + 180 + 75 + 195 = 510\,\text{cm}^2$.

Pythagoras' theorem was covered in Unit 2 of this book.

Exercise 8.2

1 Calculate the volume and surface area of each of these cylinders.

a
8 cm
6 cm

b
15 cm
3 cm

2 A large tin has a radius of 10 cm and a height of 30 cm.
A printing company prints labels to wrap around tins.

10 cm

30 cm

a What is the curved surface area of each tin?
Give your answer to the nearest whole number.

b The company charges $0.02 for each 100 cm² of labels printed.
What will be the cost of printing labels for 10 000 of these large tins?

3 A 3D object is made by attaching a square-based pyramid to one face
of a cube as shown.
Calculate the total surface area of the 3D object.

10 cm

LET'S TALK

How can you work out the area of each triangular face
of the pyramid?

12 cm

4 Two cylinders A and B are shown.

2x
h   A

x
B   2h

> Use your **specialising** skills to
> help you check which is greater by
> choosing some pairs of values for
> x and h.
> Then use your **generalising** skills to
> show which is greater for any values
> of x and h.

Cylinder B has a radius half that of A, but a height
double that of A, as shown.

a Which cylinder has the greater volume? Justify
your answer.

b Which cylinder has the greater surface area? Justify your answer.

5 A triangular prism has a cross-section in the shape of an
equilateral triangle of side length 12 cm as shown.
If the prism's length is 18 cm, calculate its total surface area.

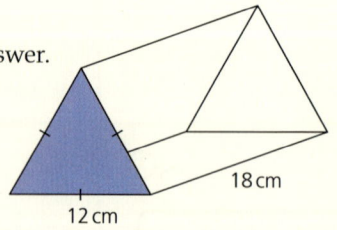

18 cm

12 cm

Now you have completed Unit 8, you may like to try the Unit 8
online knowledge test if you are using the Boost eBook.

Rational and irrational numbers

- Understand the difference between rational and irrational numbers.

Recap

In Stage 8 you studied the different types of numbers that exist, and their hierarchy. Below is a brief summary of the main points.

An **integer** is a whole number, such as 7, 40 and 626.

But an integer can also be negative, such as −18 or −86. Integers can therefore be subdivided into **positive integers** and **negative integers**.

Zero is also classified as an integer, and it is the only one that is neither positive or negative.

A **natural number** is an integer greater than zero. Natural numbers are often known as the counting numbers as they are 1, 2, 3, 4 etc.

A **rational number** is any number that can be written as a fraction. Virtually all the numbers you will have encountered up until now are rational numbers.

Using a calculator if necessary, you can check for example that 0.6 is equivalent to the fraction $\frac{6}{10}$, or that $5.7221 = 5\frac{7221}{10\,000}$.

In fact any **terminating decimal** or **recurring decimal** can be written as a fraction.

> **KEY INFORMATION**
> A terminating decimal is one which ends after a certain number of decimal places, e.g. 0.35 or 3.8267. A recurring decimal is one which repeats itself, e.g. 0.333333 etc.

You will know that if a number is a natural number it is also an integer and a rational number. The opposite, however, is not true: not every integer or rational number is a natural number.

Similarly, every integer is also a rational number but not necessarily vice versa.

There is therefore a hierarchy of numbers. Examples are shown in the Venn diagram below.

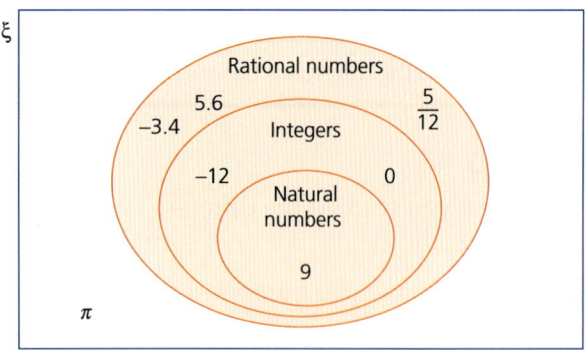

LET'S TALK
Why is π not classed as a rational number?

Irrational numbers

You can see in the Venn diagram above that π is not a rational number.

> **LET'S TALK**
> If a terminating or recurring decimal can be expressed as a fraction, what type of decimal is π?

As a rational number can always be expressed as a fraction, then it can be deduced that a number which cannot be expressed as a fraction is not rational. These numbers are known as **irrational numbers**.

There are other numbers which are irrational, such as $\sqrt{2}$ and $\sqrt{3}$.

These are clearly the square roots of two integers.

LET'S TALK
However, not all square roots of integers are irrational, e.g. $\sqrt{4}$ Why is $\sqrt{4}$ not irrational?

From your discussion above, you may have concluded that $\sqrt{4} = \pm2$ and, as both +2 and −2 are integers, then they are rational.

The square root of any positive integer will be irrational unless the positive integer is a square number.

Therefore $\sqrt{100}$ is not irrational because 100 is a square number and $\sqrt{100} = \pm10$.

Worked examples

1   Which of the following numbers are irrational? Justify your answers.

$8.\dot{3}$ 2π $\sqrt{8}$ $\sqrt{49}$

2π is irrational as it is a multiple of π and π itself is irrational.

$\sqrt{8}$ is irrational as it is the square root of a non-square number.

2   Use a calculator to work out the following multiplication and decide whether the answer is rational or irrational:

$\sqrt{3}\times\sqrt{12}$

$\sqrt{3}\times\sqrt{12}=6$

Therefore, the answer is rational.

> **KEY INFORMATION**
>
> $\sqrt{3}\times\sqrt{12}=\sqrt{36}=6$
>
> In **general**, $\sqrt{a}\times\sqrt{b}=\sqrt{a\times b}$

3   Decide whether $\sqrt{-16}$ will give a rational or irrational answer, or neither. Justify your answer.

Neither, because the square root of a number is the value that will produce that number when multiplied by itself.

No value multiplied by itself will give a negative answer, as $(-4)\times(-4)=+16$ and $4\times4=16$ as well.

Although outside the scope of this book, you may want to investigate how mathematicians represent the square roots of negative numbers.

Exercise 9.1

1 a   Which number(s) from the list below is/are rational?

$\sqrt{15}$ $\dfrac{1}{4}\pi$ $\dfrac{7}{2}$ $1.\dot{4}$

   b   Which number(s) from the list below is/are irrational?

8π $\sqrt{81}$ $0.\dot{3}\dot{2}$ 0 $\sqrt{17}$

2 The table below gives a list of numbers. Copy the table.
Decide which of the numbers are rational, which are irrational and which are neither, and then place a 'tick' in the correct column.

	Rational	Irrational	Neither
4			
$\frac{5}{8}$			
$\frac{1}{2}\pi$			
$\sqrt{9}$			
$\sqrt{-25}$			
0.333			
6.2$\overset{..}{4}$			
$\sqrt{10}$			

3 Four cards are shown below.

a Choose a card with a number that is
   i) rational   ii) irrational.
b Pick two cards which when multiplied together produce an answer that is
   i) rational   ii) irrational.
c Pick three cards which when multiplied together produce an answer that is
   i) rational   ii) irrational.

4 Without using a calculator, decide which of the following multiplications will produce a rational answer and which will produce an irrational one. Justify your answers.
a $\sqrt{8}\times\sqrt{5}$ d $\sqrt{2}\times\sqrt{128}$
b $\sqrt{8}\times\sqrt{2}$ e $\sqrt{12}\times\sqrt{16}$
c $\sqrt{10}\times\sqrt{40}$

5 For each of the following, decide whether the area of the shape is a rational or irrational number. Justify your answers.

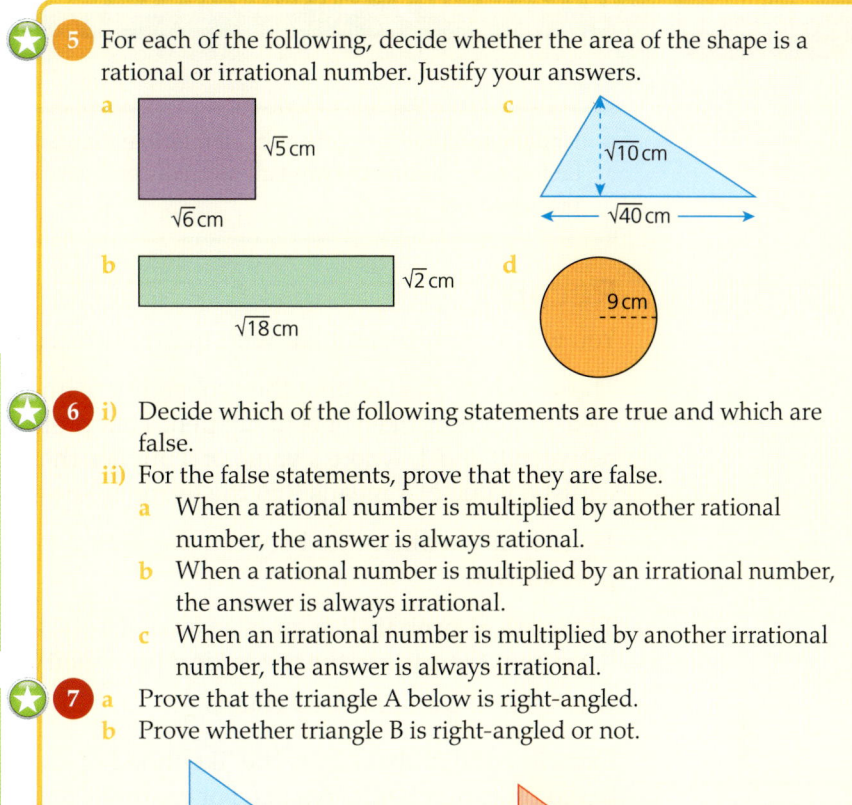

a √5 cm √6 cm

b √2 cm √18 cm

c √10 cm √40 cm

d 9 cm

KEY INFORMATION

To prove that a statement is false you can use **proof by contradiction**, i.e. find one example for which it is not true.

6 i) Decide which of the following statements are true and which are false.

ii) For the false statements, prove that they are false.

a When a rational number is multiplied by another rational number, the answer is always rational.

b When a rational number is multiplied by an irrational number, the answer is always irrational.

c When an irrational number is multiplied by another irrational number, the answer is always irrational.

KEY INFORMATION

To prove that something is correct you can use other accepted proofs in your argument.

7 a Prove that the triangle A below is right-angled.

b Prove whether triangle B is right-angled or not.

6 cm   10 cm   A   8 cm

√6 cm   √10 cm   B   √8 cm

Now you have completed Unit 9, you may like to try the Unit 9 online knowledge test if you are using the Boost eBook.

10 Mutually exclusive events

● Understand that the probability of multiple mutually exclusive events can be found by summation and all mutually exclusive events have a total probability of 1.

Recap

You will already be familiar with many of the concepts of probability.

These include the fact that the probability of an event happening is a measure of how likely it is to happen. That probability takes a value between 0 and 1, where a value of 0 implies that it is impossible, whilst a value of 1 means that the event is certain.

There is also a difference between theoretical and **experimental probability**.

Theoretical probability gives a value of how likely an event is to happen in theory, e.g. the theoretical probability of getting a head when flipping an unbiased coin is $\frac{1}{2}$.

Experimental probability is the likelihood of an event happening based on experimental results. For example, to find how likely a pin is to land point up when dropped, an experiment is conducted where a pin is dropped say 100 times. If it lands point up 60 times then the probability of landing point up is said to be $\frac{60}{100} = 0.6$.

You will also have encountered special terms associated with events. These include **independent events**, where one event happening does not affect the probability of another event happening, e.g. flipping two coins, where the result on one coin will not affect the result on the other.

Mutually exclusive events refers to those events that cannot happen at the same time, e.g. when rolling a normal six-sided dice, it is not possible to get a number which is both odd and even.

Complementary events refer to two opposing events that have a combined probability of 1, e.g. if the probability of event A happening is 0.7, then the probability of event A not happening is $1 - 0.7 = 0.3$.

> Events which are not mutually exclusive when rolling a dice would include getting a multiple of 2 or a multiple of 3. These are not mutually exclusive events as 6 is a multiple of both.

Mutually exclusive events

This unit will look at the combined probability of mutually exclusive events.

> ### Worked example
>
> A bag contains five different coloured counters: red, yellow, blue, black and green.
>
> The probability of picking each colour at random is given in the table below.
>
> | Colour | Red | Yellow | Blue | Black | Green |
> |---|---|---|---|---|---|
> | Probability | 0.4 | 0.2 | 0.1 | 0.1 | |
>
> **a** A counter is picked at random.
>
> What is the probability that it is green?
>
> The five colours are mutually exclusive as it is not possible, for example, to pick a counter that is both red and black.
>
> As picking one of the colours is certain, the sum of the probabilities must equal 1.
>
> Therefore $P(G) = 1 - 0.4 - 0.2 - 0.1 - 0.1$
>
> $\qquad\qquad = 0.2$
>
> **b** There are 300 red counters.
>
> **i** Calculate how many green counters there are.
>
> As the probability of getting a red counter is twice that of getting a green counter, then there must be twice as many reds as greens. Therefore the number of green counters is 150.
>
> **ii** Calculate how many counters there are in total.
>
> There are two ways in which this can be worked out:
> - As there are 300 red counters, the number of each of the other colours can also be worked out as in part (i).
>   That is, 300 red + 150 yellow + 75 blue + 75 black + 150 green = 750 total
> - 300 is 0.4 of the total (T). So an equation can be formed and solved:
> $$0.4 \times T = 300$$
> $$T = 300 \div 0.4$$
> $$T = 750$$

This assumes that each of the counters in the bag is equally likely to be picked.

Exercise 10.1

1 A circular spinner is divided into four different sized sectors.
Each sector is coloured a different colour.
The probability of getting each of the four colours is as follows:

Colour	Red	Blue	Yellow	White
Probability	0.3	0.4	0.2	0.2

Explain why the probability table must be incorrect.

2 Two events X and Y are mutually exclusive.
Which of the Venn diagrams below shows their relationship
accurately? Justify your choice.

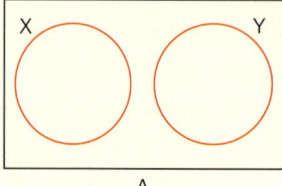
A

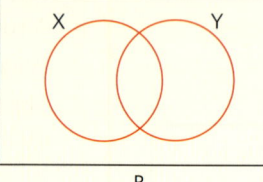
B

3 A bag contains a large quantity of counters numbered 1–5.
It is known that if a counter is picked at random then:

$P(1) = 0.24$

$P(2) = 0.36$

$P(3) = 0.12$

It is also known that picking a 5 is three times more likely than
picking a 4.

a Calculate $P(4)$.

b If there are 168 counters with a number 1, calculate how many
have a number 5.

4 A biased dice is rolled N times.
The experimental probability of getting each of the numbers 1–6 is
shown in the table below.

Number	1	2	3	4	5	6
Probability	x	0.16	0.05	0.04	0.18	0.55

a Calculate the value of x.

b If the numbers 4 and 5 were rolled a combined 88 times, calculate

   i) the number of 6s rolled

   ii) the number of times an odd number was rolled.

5 Five cards are shown.
One of the cards is picked at random.
Part of a probability table of the possible outcomes is shown below.

| 2 | 2 | 3 | 3 | 3 |

Outcome	2	3	Green	Red
Probability	0.4		0.4	

a Copy and complete the table.
b Explain why the sum of the probabilities is greater than 1.

 6 A car factory sprays its cars one of five colours.
These are red, black, silver, blue or white.
The cars are stored in the factory car park before shipping.
One day there are 500 cars in the car park and 190 of them are silver.
On the same day there are
- 20 more black cars than red ones
- 3 times as many blue cars as red ones
- 30 fewer white cars than black ones.

Assuming that each car is equally likely to be driven out of the car park first, calculate:
a the probability that the first car to be driven out is blue
b the probability that the first car to be driven out is not red.

 7 A prime number between 1 and 200 is chosen at random.
The probability table below shows the probability, to two decimal places, of picking a prime number (P) in each of the given ranges.

KEY INFORMATION
A prime number has only two factors, 1 and itself.

Number range	Probability (to 2 d.p.)
$1 \leqslant P \leqslant 40$	0.26
$41 \leqslant P \leqslant 80$	0.22
$81 \leqslant P \leqslant 120$	0.17
$121 \leqslant P \leqslant 160$	n
$161 \leqslant P \leqslant 200$	0.20

Why is your calculator not giving a whole number answer?

a What is the probability that the prime number chosen is in the range $121 \leqslant P \leqslant 160$?
b What is the probability that the prime number chosen is no bigger than 120?
c i) Write down the prime numbers in the range $1 \leqslant P \leqslant 40$.
ii) Without listing them, work out how many prime numbers are in the range $161 \leqslant P \leqslant 200$.

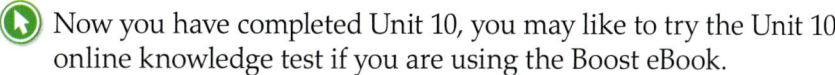 Now you have completed Unit 10, you may like to try the Unit 10 online knowledge test if you are using the Boost eBook.

Section 1 – Review

1 A cuboid has dimensions as shown.

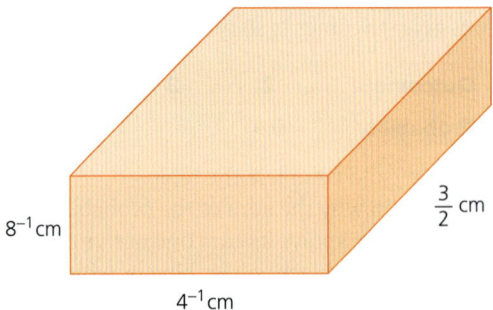

8^{-1} cm

4^{-1} cm

$\frac{3}{2}$ cm

   a Explain how the volume of the cuboid can be calculate by multiplying $\frac{1}{8} \times \frac{1}{4} \times \frac{3}{2}$

   b Calculate the surface area of the cuboid, giving your answer in the form $a \times 2^b$, where a and b are integers.

2 A square is drawn inside a circle so that the four corners of the square lie on the circumference of the circle as shown.
The diameter of the circle is 20 cm. Using Pythagoras, calculate to one decimal place the length of one side of the square.

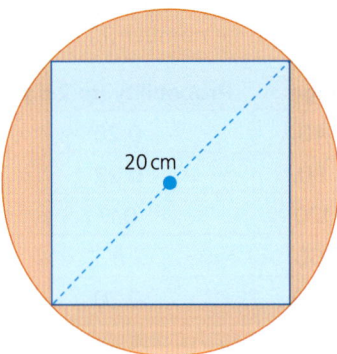

20 cm

3 You are asked to carry out a data collection exercise to see whether people's viewing habits (e.g. viewing of TV, film, news etc.) change with age.
List five things you would have to consider *before* carrying out any data collection.

4   A semicircle of radius 25 cm has another semicircle of radius 12 cm
    cut from it as shown.
    Calculate the perimeter of the remaining shape.

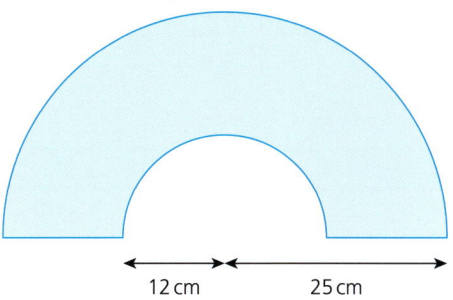

12 cm   25 cm

5   Evaluate the following.
    a $20 + (12 - 7)^2 (8 + 2)$
    b $\left(\dfrac{30+3}{3}\right)^2 - 4^2 \left(\dfrac{8-2}{4}\right)$

6   A two and a half hour high definition film uses up 1.45 GB of
    storage on a computer's hard drive.
    If someone is able to download the film at a rate of 42 MB per
    second, calculate how long it will take them to download the
    whole film.

7   Give an example, and a justification for, two sets of data which
    are likely to have:
    a   a negative correlation
    b   a positive correlation
    c   no correlation.

8   A triangular prism is shown.

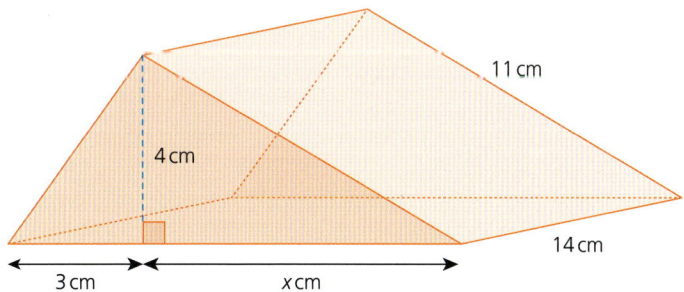

11 cm

4 cm

14 cm

3 cm x cm

    a   Calculate the value of x.
    b   Calculate the total surface area of the prism.
    c   Calculate the volume of the prism.

9 Deduce whether the area of each of the following shapes is rational or irrational.

a

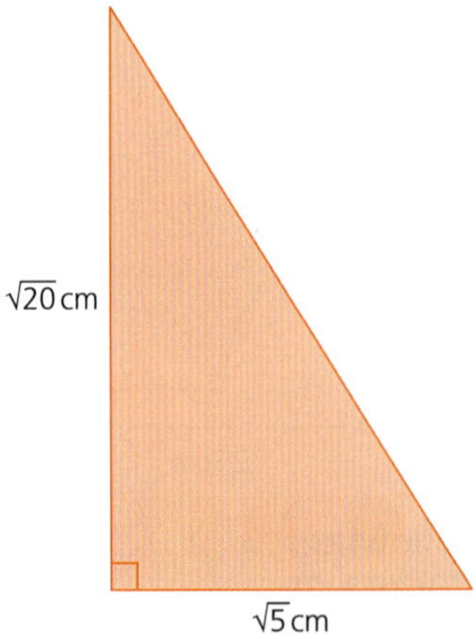

$\sqrt{20}$ cm

$\sqrt{5}$ cm

b

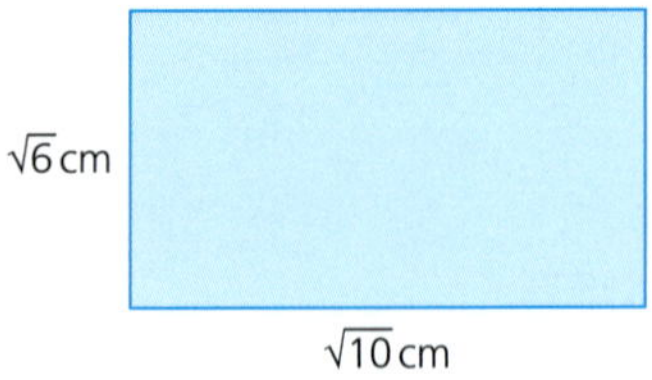

$\sqrt{6}$ cm

$\sqrt{10}$ cm

10 A student wished to work out the experimental probability of girls in her school liking mathematics.

She asked all 320 girls whether they like mathematics.

280 of them said they did.

a   Are the outcomes 'being a girl' and 'liking mathematics' mutually exclusive? Justify your answer.

b   Are the outcomes 'not liking mathematics' and 'liking mathematics' mutually exclusive? Justify your answer.

SECTION 2

History of mathematics — Sir Isaac Newton

"No matter how correct a mathematical theorem may appear to be, one ought never to be satisfied that there was not something imperfect about it until it also gives the impression of being beautiful."

George Boole

▲ Isaac Newton

Isaac Newton was born in Lincolnshire in England and was possibly the greatest scientist and mathematician ever to have lived. He was 22, and on leave from university (isolating in the countryside due to the bubonic plague) when he began work in mathematics, optics, dynamics, thermodynamics, acoustics and astronomy. Newton studied gravitation and the idea that white light is a mixture of all the rainbow's colours. Newton also designed the first reflecting telescope and the sextant. Newton is widely regarded as the 'father of calculus', which is the mathematics of continuous changes.

In 1687 Newton published *Philosophiae Naturalis Principia Mathematica*, one of the greatest scientific books ever written. The movement of the planets was not understood before Newton proposed the Laws of Motion and the Law of Universal Gravitation. The idea that the Earth rotated about the Sun was introduced in India, Arabia and ancient Greece, but Newton explained why it did so.

11 Rounding and estimating numbers

- Understand that when a number is rounded there are upper and lower limits for the original number.
- Multiply and divide integers and decimals by 10 to the power of any positive or negative number.
- Estimate, multiply and divide decimals by integers and decimals.

LET'S TALK

Does that mean there are exactly 25 000 spectators? What are the maximum and minimum numbers of spectators that there could be?

A number line is drawn with numbers going up in 1000's.

KEY INFORMATION

The fact that 24 500 is included in the range but 25 500 is not is shown on the number line using different circles. A solid circle implies the number is included, whilst a hollow circle implies it is not.

Upper and lower limits

When describing the attendance at a sporting event a commentator might say 'Today there are 25 000 spectators'.

To have some idea of the possible number of people who were in attendance we need to know how the number has been rounded.

If the number has been rounded to the nearest 1000, then the following number line shows the range of possible values.

24 000   25 000   26 000

Any number in the highlighted region would be rounded to 25 000. The start and end points of the region are halfway between the numbers.

The problem is what happens if a number is exactly halfway, e.g. 24 500 or 25 500?

As with other topics involving rounding, such as decimal places and significant figures, numbers which are halfway get rounded up, i.e. in this case a number of 24 500 would get rounded up to 25 000, and 25 500 would get rounded up to 26 000.

As an inequality the number range (N) can be expressed as follows:

$$24\,500 \leqslant N < 25\,500$$

24 500 is known as the **lower limit**, whilst 25 500 is known as the **upper limit**.

KEY INFORMATION

Note that, although technically 25 500 would be rounded up to 26 000, it is still regarded as the upper limit of this range of possible numbers.

> A number line is drawn with numbers going up in 5000's.

If the number of spectators had been rounded to the nearest 5000, then the range of the possible numbers (N) of spectators would be different.

As shown above, the lower limit for the number of spectators would now be 22 500 and the upper limit 27 500.

$$22\,500 \leqslant N < 27\,500$$

Worked examples

1 A sweet manufacturer sells tubs of sweets. They state that each tub contains 200 sweets.

Give the range of the number of sweets (S) as an inequality, if the number of sweets stated on the tub is rounded to:

a the nearest 50 sweets

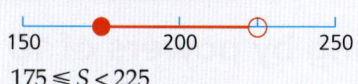

$$175 \leqslant S < 225$$

b the nearest 10 sweets.

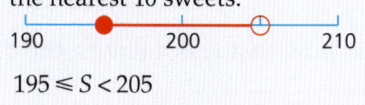

$$195 \leqslant S < 205$$

2 The number of fish in a large aquarium is given as 3500.

If the number has been given to the nearest 50, calculate both the upper and lower limits for the number of fish in the aquarium.

Lower limit = 3475
Upper limit = 3525

> A number line always helps to visualise the problem.

Exercise 11.1

1 Give the upper and lower limits for the following approximations.
  a i) 5000 rounded to the nearest 1000   ii) 5000 rounded to the nearest 500
  b i) 800 rounded to the nearest 50   ii) 800 rounded to the nearest 20
  c i) 10 000 rounded to the nearest 100   ii) 10 000 rounded to the nearest 10
  d i) −400 rounded to the nearest 100   ii) −400 rounded to the nearest 10
  e i) 1 rounded to the nearest 1   ii) 1 rounded to the nearest 0.5

2 A conference centre is told that they have approximately 2500 people attending.
  a The number attending the conference is a figure rounded to the nearest 500.
     The organisers set out 2700 seats.
     Give a **convincing** argument to explain whether this is enough seats.
  b If the number of 2500 attending the conference is a figure rounded to the nearest 50, explain why setting out only 2480 seats could be enough.

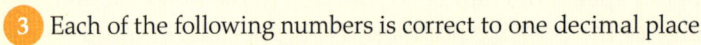

3   Each of the following numbers is correct to one decimal place.
   i)   Give the lower and upper limits in each case.
   ii) Using x as the number, express the range in which the number lies as an inequality.
       a   3.5   b   18.1   c   9.0   d   50.0   e   0.2

4   At a school sports day, the winning time for the 200 m race is given as 25.6 seconds to 1 d.p.
   a   Illustrate the lower and upper limits of the time on a number line.
   b   Using T seconds for the time, express the range of values in which T must lie as an inequality.
   c   The school athletics records book gives the current record for the 200 m as 25.65 seconds.
       Explain whether the new recorded time of 25.6 seconds breaks the existing record.

5   Each of the following numbers is correct to two significant figures.
   i)   Using x as the number, express the range in which the number lies as an inequality.
   ii) Write a number to three significant figures that falls within this range.
       a   6.8   b   0.85   c   640   d   8000   e   0.028   f   10 000

6   A farmer measures the dimensions of his rectangular field to two significant figures.
   He records the length as 450 m and the width as 200 m.
   He has sheep to put in the field.
   The regulations state that the sheep need at least 87 000 m² of field.
   Can he legally put his sheep in the field? Justify your answer thoroughly.

Multiplying and dividing by powers of 10

You will already be familiar with multiplying and dividing by powers of 10.

However, this section will look at how that is done with the different ways of writing powers of 10.

For example, 10^{-2} can be written as $\dfrac{1}{10^2}$ and therefore as $\dfrac{1}{100}$ or 0.01.

Therefore, a multiplication by 10^{-2} is the same as multiplying by $\dfrac{1}{100}$ or 0.01. Similarly, a multiplication by $\dfrac{1}{100}$ is equivalent to dividing by 100.

Worked examples

1   Without using a calculator, work out 5.6×10^{-2}.

   Method 1: $5.6 \times \dfrac{1}{100} = 0.056$

   Method 2: $5.6 \times 0.01 = 0.056$

   Method 3: $5.6 \div 100 = \dfrac{5.6}{100} = 0.056$

> Depending on the numbers used and the operation, you may find one method easier than another. It is therefore useful to be confident with the different methods.

> Multiplication does not always produce an answer bigger than the original number. $10^{-2} = 0.01$ is a number between 0 and 1. Multiplying by a number between 0 and 1 produces an answer that is smaller than the original number.

2 Without using a calculator, evaluate the division $4.8 \div 10^{-3}$.

Method 1: $4.8 \div \dfrac{1}{10^3} = 4.8 \div \dfrac{1}{1000} = 4800$

Method 2: $4.8 \div 0.001 = 4800$

Method 3:   Division by a fraction is the same
as multiplying by the reciprocal of that fraction.

Therefore $4.8 \div \dfrac{1}{1000} = 4.8 \times 1000 = 4800$

> Division does not always produce an answer smaller than the original number. $10^{-3} = 0.001$ is a number between 0 and 1. Dividing by a number between 0 and 1 produces an answer that is bigger than the original number.

LET'S TALK

Look at the worked example and discuss how the three methods for each part are in fact the same calculation but written in different formats.

Exercise 11.2

1 Here are seven cards with a mathematical calculation.
   a Identify the odd one out.
   b Give a reasoned argument for your choice in part (a).

| Multiply by 10^3 | Divide by 0.001 | Divide by $\dfrac{1}{10^3}$ | Divide by 10^{-3} |

| Divide by $\dfrac{1}{1000}$ | Divide by $\dfrac{1}{10^{-3}}$ | Multiply by 1000 |

2 Eight calculations are given below.
   Identify which would give an answer bigger and which smaller, than the number they are applied to.
   a Multiply by 10^2
   b Multiply by 10^{-2}
   c Divide by $\dfrac{1}{10^3}$
   d Divide by 10^{-1}
   e Multiply by $\dfrac{1}{10^{-2}}$
   f Multiply by $\dfrac{1}{10^3}$
   g Divide by 10^2
   h Divide by $\dfrac{1}{10^{-4}}$

3 Work out the answers to the following calculations.
   a 7.0×10^2
   b $12 \div 10^3$
   c $8.6 \div 10^2$
   d 0.58×10^0
   e 101×10^{-2}
   f $-98.7 \div 10^{-3}$
   g $0.082 \div 10^{-5}$
   h -0.09×10^{-1}

4 The calculation 0.18×10^4 gives the same answer as $0.18 \times \dfrac{1}{10^{-4}}$ and $0.18 \div 10^{-4}$.

   i) In each of the following write an equivalent calculation by changing either the operation, the power of 10 or both.
   ii) Write down the answer to the calculation.
      a 12×10^2
      b 15.7×10^5
      c 0.06×10^{-1}
      d $4.87 \div 10^{-2}$
      e $0.9 \div 10^3$
      f $0.07 \div 10^{-6}$

5 A large square has a side length of 12 cm. A small square has a side length of 0.1 cm.
  a i) Write down a calculation, using powers of 10, to work out how many of the smaller squares fit into the larger square.
     ii) Is there another calculation which would give the same answer? Justify your answer.
  b How many of the smaller squares fit in the larger square?

6 A large cube has an edge length of 8 cm. A small cube has an edge length of 0.01 cm.
  a i) Using powers of 10 in the answer, write down a possible calculation to work out how many of the smaller cubes fit inside the larger one.
     ii) Work out the answer to your calculation in part (i).
  b i) Using powers of 10 in the answer, write down a possible calculation to work out how many times smaller the surface area of the small cube is, compared with the surface area of the large cube.
     ii) Work out the answer to your calculation in part (i).

7 For each pair of calculations below, find the value of x in order to make:
  i) the second calculation 100 times bigger than the first
  ii) the second calculation 10 times smaller than the first.

  a 7.5×10^2 7.5×10^x
  b 4×10^{-1} $4 \div 10^x$
  c $0.111 \div 10^{-4}$ 0.111×10^x
  d $12.6 \div 10^3$ 12.6×10^{-x}
  e 0.1×10^{-6} $0.1 \div 10^{-x}$

Further multiplication and division

In Stage 8 you studied how to multiply decimals and integers by decimals. This section will extend that to include division by integers and decimals too.

The methods for multiplication and division that you have covered before are still applicable here. The focus will therefore be on developing and understanding the different methods and selecting which are the most appropriate for the given situation.

In Stage 7 you were introduced to the method of long multiplication.

For example, multiply 182 × 36.
● Arrange the numbers above each other, such that the units line up.

	Hundreds	Tens	Units
	1	8	2
×		3	6

- Multiply each of the digits in 182 by the 6.

$$
\begin{array}{ccc}
 & 1 & 8 & 2 \\
\times & & 3 & 6 \\
\hline
1 & {}_4 0 & {}_1 9 & 2 \\
\end{array}
$$

- Multiply each of the digits in 182 by the 3 (which is 3 tens, i.e. 30).

$$
\begin{array}{cccc}
 & 1 & 8 & 2 \\
\times & & 3 & 6 \\
\hline
1 & 0 & 9 & 2 \\
{}_2 5 & 4 & 6 & 0 \\
\end{array}
$$

> The zero in the units column indicates that 182 was multiplied by 30 rather than by 3.

- Add the two answers together.

$$
\begin{array}{cccc}
 & 1 & 8 & 2 \\
\times & & 3 & 6 \\
\hline
1 & 0 & 9 & 2 \\
5 & 4 & 6 & 0 \\
\hline
6 & {}_1 5 & 5 & 2 \\
\end{array}
$$

Therefore $182 \times 36 = 6552$

This can be used to work out, say, 18.2×3.6 or 1.82×360.

Worked example

a   Given that $182 \times 36 = 6552$, evaluate 18.2×3.6.

We know that $18.2 = 182 \times \dfrac{1}{10}$

We also know that $3.6 = 36 \times \dfrac{1}{10}$

Therefore, 18.2×3.6 can be written as $182 \times \dfrac{1}{10} \times 36 \times \dfrac{1}{10}$

> **KEY INFORMATION**
> 'Commutative' means that it can be done in any order.

As multiplication is commutative, $182 \times \dfrac{1}{10} \times 36 \times \dfrac{1}{10}$ can be rearranged as:

$182 \times 36 \times \dfrac{1}{10} \times \dfrac{1}{10} = 182 \times 36 \times \dfrac{1}{100}$

$\qquad = 6552 \times \dfrac{1}{100}$

$\qquad = 65.52$

> Each number is 10 times smaller than the original numbers, therefore the answer is 100 times smaller.

b Given that $182 \times 36 = 6552$, evaluate 1.82×360.

We know that $1.82 = 182 \times \dfrac{1}{100}$

We also know that $360 = 36 \times 10$

Therefore, 1.82×360 can be written as $182 \times \dfrac{1}{100} \times 36 \times 10$

$182 \times \dfrac{1}{100} \times 36 \times 10$ can be rearranged as:

$$182 \times 36 \times 10 \times \dfrac{1}{100} = 65\,520 \times \dfrac{1}{100}$$
$$= 655.2$$

> As one number was 100 times smaller than the original number and the other was 10 times bigger than the original number, the answer will be 10 times smaller.

c By estimation, show that the answers in parts (a) and (b) are correct.

18.2×3.6 can be estimated using the rounded values $18 \times 4 = 72$

Therefore, the answer of 65.52 is correct.

1.82×360 can be estimated using the rounded values $2 \times 350 = 700$

Therefore, the answer of 655.2 is correct.

> When manipulating numbers by factors of 10, it is easy to make careless mistakes, therefore being able to estimate the answer is important.

Another method introduced for doing multiplication with pen and paper was to use the grid method. To multiply 182×36 again:

- 182 is a three-digit number.
  36 is a two-digit number.
  Draw a 3 × 2 grid and draw diagonals as shown.

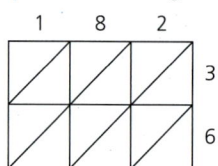

- Multiply each digit of 182 by each digit of 36 and enter each answer in the corresponding part of the grid.

● Extend the diagonals.

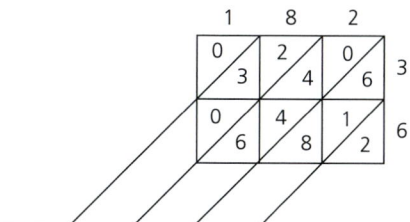

● Add all of the digits in each diagonal together, working from right to left.

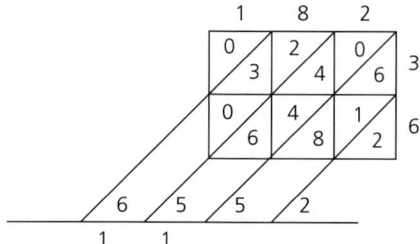

The answer, as before, is 6552.

The grid can be adapted for decimal multiplication too.

For the multiplication 18.2×3.6 the decimal point can be placed as shown below.

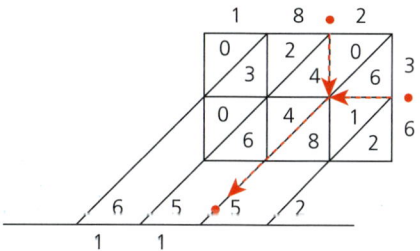

Follow the decimal points along the grid lines until they meet, and then follow that diagonal, to give the position of the decimal point in the answer, i.e. 65.52.

With division, you will be familiar with the following method.

Calculate 576 ÷ 18.

In other words, how many times does 18 go into 576?
- Set up the division as follows:

$$18 \overline{\smash{\big)}\ 5 \quad 7 \quad 6}$$

- Working from left to right, work out how many times 18 goes into each digit, i.e. firstly, how many times does 18 go into 5?

$$\begin{array}{c} 0 \\ 18 \overline{\smash{\big)}\ 5 \quad {}^5 7 \quad 6} \end{array}$$ *0 times therefore with 5 left over as the **remainder** and we move it to the next column to make 5.*

- How many times does 18 go into 57?

$$\begin{array}{c} 0 \qquad 3 \\ 18 \overline{\smash{\big)}\ 5 \quad {}^5 7 \quad {}^3 6} \end{array}$$ *3 times with 3 remainder*

- How many times does 18 go into 36?

$$\begin{array}{c} 0 \qquad 3 \qquad 2 \\ 18 \overline{\smash{\big)}\ 5 \quad {}^5 7 \quad {}^3 6} \end{array}$$ *twice with 0 remainder*

Therefore 576 ÷ 18 = 32

The answer to this calculation can be used to calculate the answers to other problems, e.g. 5.76 ÷ 1.8

Both of the original numbers have been multiplied or divided by multiples of 10 to give these two numbers.

Writing them in terms of the original numbers and powers of 10 gives the following calculation:

$$\frac{576}{100} \div \frac{18}{10}$$

As a division by a fraction is equivalent to multiplying by its reciprocal, the above calculation can be written as:

$$\frac{576}{100} \times \frac{10}{18} = \frac{576}{18} \times \frac{10}{100}$$
$$= 32 \times \frac{1}{10}$$
$$= 3.2$$

LET'S TALK

There are many other ways of carrying out this calculation. Discuss some possible alternative methods and decide which you find easiest to use.

Exercise 11.3

1 For each of the following calculations:
  i) work out an estimate for the answer
  ii) calculate the answer.
  a 48.2×12
  b 5.11×36
  c -56×10.4
  d $96 \div 4.4$
  e $427 \div (-6.6)$
  f $-52.8 \div 48$

2 A coil of metal wire with a total length of 324 m is cut into pieces each of length 2.7 m.
  a How many pieces can be cut from the coil?
  b **i)** Is there any wire left over? If so, how much?
    ii) Justify your answer to part (i).

3 A 98.6 m length of material is to be shared equally between 12 people. Calculate to the nearest cm how much each person will receive.

4 For each of the following statements:
  i) decide whether the statement is true or false
  ii) give a reasoned justification for your answer.
  a $92.6 \times 41 = 9.26 \times 410$
  b $-6.04 \times 1.2 = 604 \times (-0.012)$
  c $55 \div 6.2 = 550 \div 0.62$
  d $88.2 \div 106.2 = \dfrac{8.82}{10.62}$

5 A sheet of metal has dimensions as shown.

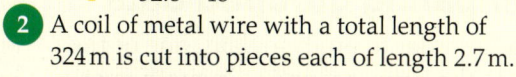

112.5 cm

27.8 cm

Squares of side length 4.5 cm are cut from it.
  a What is the maximum number of squares that can be cut from the metal sheet?
  b What is the area of metal wasted, once the maximum number of squares have been cut from it?

6 A car's petrol tank can hold a maximum of 72.7 litres of petrol.
The car can travel an average of 24.2 km per litre of petrol.
The cost of petrol is $1.23 per litre.
  a **i)** Estimate the cost of travelling a distance of 1200 km.
    ii) By refining your figures, work out an **improved** estimate for travelling the 1200 km distance.
  b Calculate the cost of travelling the 1200 km journey.

Now you have completed Unit 11, you may like to try the Unit 11 online knowledge test if you are using the Boost eBook.

Further data interpretation

- Use mode, median, mean and range to compare two distributions, including grouped data.
- Interpret data, identifying patterns, trends and relationships, within and between data sets, to answer statistical questions. Make informal inferences and **generalisations**, identifying wrong or misleading information.

Analysing and comparing grouped data

LET'S TALK
Discuss the advantages and disadvantages of each measure of 'average'.

You will already know that in order to analyse data, one of the methods used is to calculate the 'average' value. In mathematics, the word 'average' can imply one of three different calculations: the **mean**, the **median** and the **mode**.

The range is also often used as it gives an idea of how spread out the data is.

Worked example

The shoe sizes of a class of 29 students are shown in the frequency table below.

Shoe size	3	4	5	6	7	8	9
Frequency	1	2	5	7	7	4	3

Calculate the mean, median, mode and range of the shoe sizes.

KEY INFORMATION
Note that although the shoe sizes themselves are integer values, the mean does not have to be.

$$\text{Mean} = \frac{(3\times1)+(4\times2)+(5\times5)+(6\times7)+(7\times7)+(8\times4)+(9\times3)}{29} = 6.41 \text{ (to 2 d.p.)}$$

The median is the middle number when the values are arranged in order.

As there are 29 students, the middle value will be the 15th one (14 below and 14 above).

The shoe sizes are already arranged in order in the table, therefore the 15th value can be found by first calculating the **cumulative frequencies**, i.e.

Shoe size	3	4	5	6	7	8	9
Frequency	1	2	5	7	7	4	3
Cumulative frequency	1	3	8	15	22	26	29

KEY INFORMATION
A cumulative frequency represents the total frequency up to that point and is simply the sum of the frequencies to that point.

The 15th value falls in the category of shoe size 6.

Therefore the median shoe size is 6.

There are two modal values, shoe sizes 6 and 7, as they are both more common than any other shoe size.

The range of shoe sizes is 9 – 3 = 6. This means there is a difference of 6 shoe sizes between the largest and smallest shoe sizes.

> If there is an odd number of data values then there will be one middle value. If there is an even number of data values then there will be a middle pair. The median is calculated by working out the mean of the data values for that middle pair.

KEY INFORMATION

A common error is to work out the difference between the highest and lowest frequency values.

Note how the group sizes are all the same and that there is no overlap between the groups.

Grouped data are not as straightforward because the data values are put into groups and therefore the calculations for the mean, median, mode and range are not exact, but are instead an approximation.

Assume that the shoe size data given above had instead been presented as grouped data.

One possible grouping is shown below.

Grouped shoe size	3–4	5–6	7–8	9–10
Frequency	3	12	11	3

In the following calculations, assume that we no longer know the individual results displayed in the original table.

Calculating an accurate mean from this table of data is now impossible as we don't know where, within each group, the individual data values lie.

To estimate the mean we assume that each value within the group takes the **mid-interval value** which, as the name implies, is the value in the middle of the group, i.e.

LET'S TALK

Here the estimate of 6.47 is close to the actual mean of 6.41 calculated earlier. What factors would make the estimate less accurate?

Grouped shoe size	3–4	5–6	7–8	9–10
Mid-interval value	3.5	5.5	7.5	9.5
Frequency	3	12	11	3

$$\text{Mean} = \frac{(3.5 \times 3) + (5.5 \times 12) + (7.5 \times 11) + (9.5 \times 3)}{29} = 6.47 \text{ (to 2 d.p.)}$$

LET'S TALK

Although beyond the syllabus, there are methods for estimating a value for the median. How could you estimate where within the 5–6 group the 15th student is likely to be?

The median too must be an estimate as the individual values are not known in this table.

The median value is still the 15th value, which, worked out cumulatively as before, falls in the 5–6 interval.

The median is therefore given as the interval 5–6.

As the individual data is not known, the modal class is given, i.e. the modal class is 5–6 as this is the class interval with the highest frequency.

Notice that when the individual data were known, the mode was both 6 and 7, whilst when grouped the modal class is 5–6. Therefore it is possible for individual modal values (i.e. 7) to fall outside the modal class (i.e. 5–6). The way in which the data are grouped may also affect the modal class.

KEY INFORMATION

Another term for **modal class interval** is the modal group.

The range will also be an estimate from grouped data as once again the individual values are not known. It is therefore assumed that the smallest value takes the number from the bottom of the lowest group (in this case 3) whilst the greatest value takes the number from the top of the highest group (in this case 10).

$$\text{Range} = 10 - 3 = 7$$

In order to compare the accuracy of these estimates, look at the comparison table below.

LET'S TALK

Are the estimates good?

Give reasoned arguments for your decisions.

	Individual data (actual values)	Grouped data (estimated values)
Mean	6.41	6.47
Median	6	5–6
Mode	6 and 7	5–6
Range	6	7

Worked example

The age distribution (A) of people working in a supermarket is as shown below.

Age	Frequency
$15 \leqslant A < 25$	8
$25 \leqslant A < 35$	21
$35 \leqslant A < 45$	30
$45 \leqslant A < 55$	28
$55 \leqslant A < 65$	17
$65 \leqslant A < 75$	6

KEY INFORMATION

$15 \leqslant A < 25$ is an inequality to represent ages from 15 up to but not including 25.

Calculate an estimate for the **mean**, **median**, **mode** and **range** of the ages of the supermarket's workforce. Assume age is **continuous data**.

To calculate the mean, calculate the **mid-interval values**. The table is therefore:

Age	Mid-interval value	Frequency
$15 \leqslant A < 25$	20	8
$25 \leqslant A < 35$	30	21
$35 \leqslant A < 45$	40	30
$45 \leqslant A < 55$	50	28
$55 \leqslant A < 65$	60	17
$65 \leqslant A < 75$	70	6

KEY INFORMATION

Care must be taken when calculating the **mid-interval value**. Here the data are presented as **continuous data** therefore the **mid-interval values** are 20, 30, 40 etc. If the data had been presented as discrete, e.g. only being **integer** values in the ranges 15–24, 25–34 etc. the **mid-interval values** would be 19.5, 29.5 etc.

$$\text{Mean} = \frac{(20 \times 8) + (30 \times 21) + (40 \times 30) + (50 \times 28) + (60 \times 17) + (70 \times 6)}{110} = 43.9 \text{ years old}$$

As there are 110 people working in the supermarket, the median value is the mean of the 55th and 56th values.

The cumulative frequency column, added to the table below, shows that both the 55th and 56th values fall in the $35 \leqslant A < 45$ class interval.

Age	Frequency	Cumulative frequency
$15 \leqslant A < 25$	8	8
$25 \leqslant A < 35$	21	29
$35 \leqslant A < 45$	30	59
$45 \leqslant A < 55$	28	87
$55 \leqslant A < 65$	17	104
$65 \leqslant A < 75$	6	110

So the **median** class interval is $35 \leqslant A < 45$.

The **modal** class interval is also $35 \leqslant A < 45$ as this is the group with the greatest frequency.

The estimate for the range is the difference between the potential youngest age and the potential oldest age.

Lower limit is 15.

Upper limit is 75.

The range is $75 - 15 = 60$ years.

> Remember that with $65 \leqslant A < 75$, 75 is regarded as the upper limit.

LET'S TALK

In many countries, the youngest age someone can work full-time is 16. How might this affect the results? How could the class intervals be changed to take this into account?

Exercise 12.1

1 The times (t minutes) taken for a class of students to complete a mathematics exercise are given in the table below.

Time (min)	$0 \leqslant t < 5$	$5 \leqslant t < 10$	$10 \leqslant t < 15$	$15 \leqslant t < 20$	$20 \leqslant t < 25$
Frequency	1	3	12	7	6

a Explain whether the time, t, is discrete or continuous data.

b The teacher expects the average time it takes for a student to complete the exercise to be under 13 minutes.

i) Is the teacher's prediction correct?

ii) Give a reasoned justification for your answer to part (i).

2 As part of a fundraising event, two classes of students enter a cross-country race. Their times (T minutes) for completing the race are shown in the table.

Time (min)	Class A	Class B
$10 \leqslant T < 20$	1	0
$20 \leqslant T < 30$	3	11
$30 \leqslant T < 40$	8	5
$40 \leqslant T < 50$	10	2
$50 \leqslant T < 60$	4	12
$60 \leqslant T < 70$	2	0

LET'S TALK

In this table the inequality $10 < T \leqslant 20$ is used instead of $10 \leqslant T < 20$. Does this affect calculations for continuous data? Would it affect calculations for discrete data?

a Compare the performances of the two classes for the race, by considering the mean, median, mode and range of their data.

b Which class did better at the cross-country race? Write a short report to give a reasoned answer to this question.

3 For each of the following statements:
  i) state whether the statement is true or false
  ii) if it is false, give a counter example as proof.
  a The individual mean result is never the same as the grouped mean result.
  b The individual median value will always fall within the grouped median range.
  c The mode of the individual data will always fall within the grouped modal class interval.
  d The range of the individual data is always less than the range for when the data are grouped.

 4 Each of the following sets of grouped continuous data has one frequency value missing, labelled N.
  i) Decide on a possible value for N which satisfies the condition stated.
  ii) Justify your choice of value.

  a The estimated modal class is $20 < x \leqslant 25$.

Class interval	Frequency
$5 < x \leqslant 10$	3
$10 < x \leqslant 15$	N
$15 < x \leqslant 20$	8
$20 < x \leqslant 25$	10
$25 < x \leqslant 30$	4

  c The estimated median class interval is $15 < x \leqslant 20$.

Class interval	Frequency
$5 < x \leqslant 10$	3
$10 < x \leqslant 15$	18
$15 < x \leqslant 20$	8
$20 < x \leqslant 25$	N
$25 < x \leqslant 30$	4

  b The estimated range is 20.

Class interval	Frequency
$5 < x \leqslant 10$	N
$10 < x \leqslant 15$	12
$15 < x \leqslant 20$	8
$20 < x \leqslant 25$	10
$25 < x \leqslant 30$	4

  d The estimated mean is 18.

Class interval	Frequency
$5 < x \leqslant 10$	3
$10 < x \leqslant 15$	6
$15 < x \leqslant 20$	8
$20 < x \leqslant 25$	11
$25 < x \leqslant 30$	N

> **LET'S TALK**
>
> Is there more than one possible value for N in each part?
> If so, what are they and why?
> If not, why not?

Interpreting graphs and 'fake news'

Statistics and graphs are an important part of modern-day communication. To be able to analyse and interpret data and to present them clearly is vital.

As data and their presentation are so critical, they can sometimes be used to mislead the reader. There is an important difference between data which are factually incorrect and data which are misleading. If data are factually incorrect and yet presented as the truth, then that may be an attempt to mislead the reader simply by dishonesty. Misleading data, in contrast, are often factually correct but are presented in such a way that unsuspecting readers will interpret them incorrectly, which is often the intention.

This section will focus on the second sorts; in other words on data that are correct but which are presented or interpreted in a misleading way.

Worked examples

1. This is a graph printed in *The Daily Insight* newspaper comparing its sales against one of its competitors, *In-Fact*.

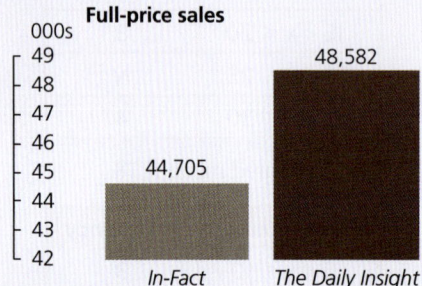

a Assuming the figures are correct, give two reasons why the graph may be misleading.
  ● The bar for *The Daily Insight* appears to be more than double the height of the bar for *In-Fact*. However, the vertical axis does not start at zero.
  ● The difference between the two is approximately 4000 out of approximately 45 000 sales.
  ● This graph is just for full-price sales. It ignores other types of sales, e.g. discounted sales, online sales, membership viewings etc., which may impact the relative shapes of the graph.

LET'S TALK
What is the percentage difference in the sales? Is this figure less misleading?

b   Redraw the graph so that it is not as misleading.

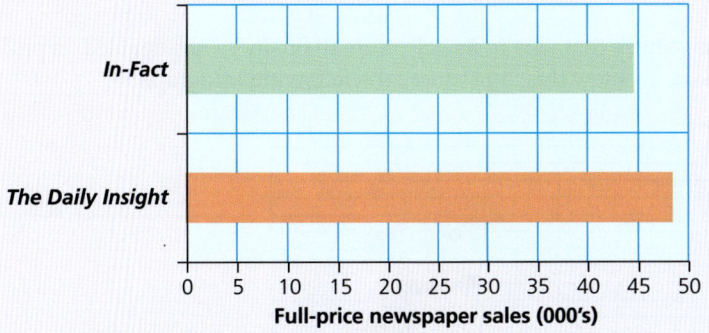

LET'S TALK

Would the timeframe for this data collection be useful to know? Justify your answer.

2   Two shops next to each other sell different goods. One sells ice-creams, the other woollen jumpers. For each month over the period of one year, they both work out the mean daily sales of their items. A scatter graph of the data is shown below.

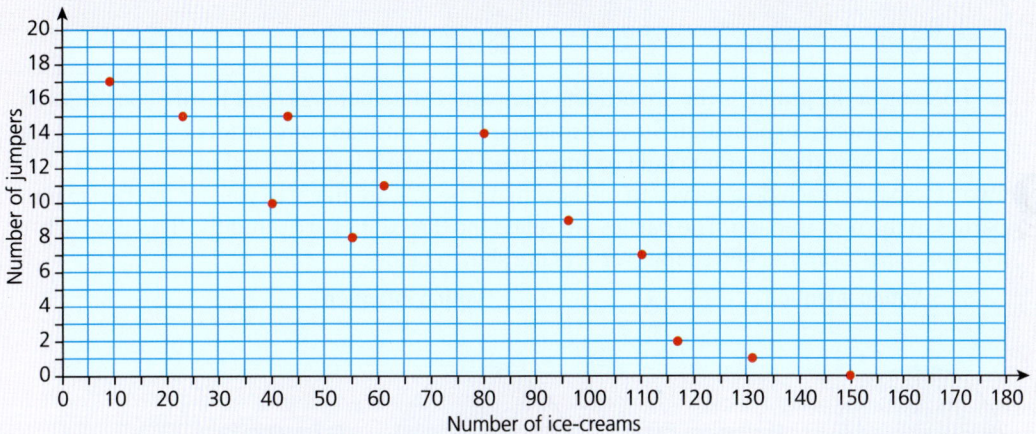

Looking at these data, the ice-cream vendor says that an increase in the sale of ice-creams leads to a decrease in the sale of jumpers.

a   Explain why this conclusion is not likely to be correct.

Although there is a negative correlation between the two variables, that does not mean that one variable causes the other.

They may be related due to another third variable.

b   What other variable may be causing the apparent correlation between ice-cream sales and jumper sales?

The likeliest variable to be affecting both of these variables is the outdoor temperature. On a hot day we would expect to see a large number of ice-creams sold and a small number of jumpers sold. On a cold day, the opposite is likely to be the case.

KEY INFORMATION

A correlation does not imply a **causal effect**, i.e. one variable does not necessarily affect the other. This is known as a **spurious correlation**.

Exercise 12.2

⭐ **1** An advertising company states that, due to its help, a small business has doubled its profits from one year to the next. To show this, it produces the following infographic.

a Looking at the graph, is the advertising company's statement correct? Justify your answer.

b Give a reasoned explanation for why this graph may mislead some readers.

c Describe how the diagram could be **improved** to make it less misleading.

⭐ **2** Two newspapers each do a survey to see how many people are likely to vote for the two main parties in the next election. The parties are called the 'Freedom Party' and the 'Popular Party'. The pie charts from the newspaper surveys are shown below.

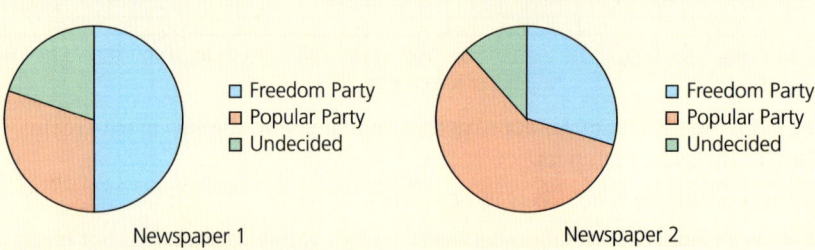

For each statement in parts (a)–(c):

i) decide whether the statement is true or false or whether it is not possible to tell

ii) justify your choice in part (i).

a The same number of people were surveyed by newspaper 1 and newspaper 2.

b The number of people choosing 'Freedom Party' in the newspaper 1 survey was more than the number of people choosing 'Freedom Party' in the newspaper 2 survey.

c A smaller proportion of people were undecided in the newspaper 2 survey compared with the newspaper 1 survey.

d Name two factors that would affect the reliability of the surveys.

3 a Carry out a brief survey of your choice.
  b Present the results of your survey in a misleading way.
  c Present the results in a non-misleading way.

4 In 2020, the official world record for the marathon was held by the Kenyan Eliud Kipchoge, with a time of 2:01:39.
An athlete at a local club wants to be the world record holder.
He enters a marathon race each year and records his time. A graph showing his times is given below.

LET'S TALK
Discuss the misleading aspects of your survey and graph with another student.

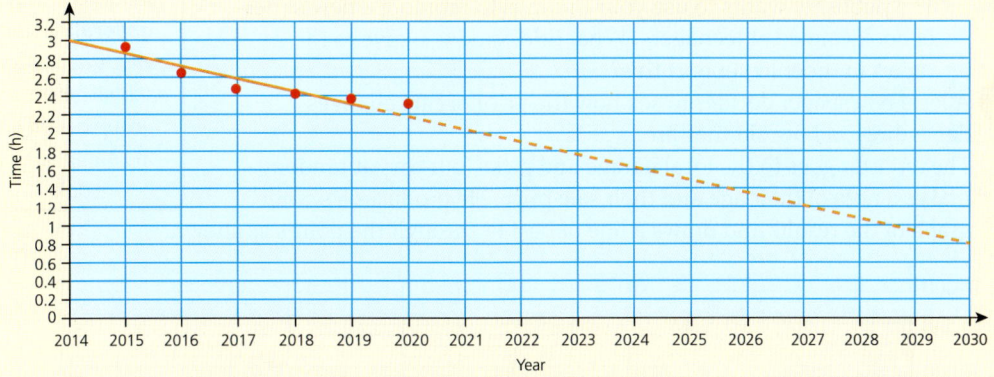

In 2020 he records his fastest time and the local paper publishes the graph.
The newspaper states that the runner will beat the world record in 2022 and, incredibly, that he will run a marathon in 1 hour in 2030.
  a What assumptions has the newspaper made for these predictions?
  b What would be a more accurate deduction from the graph?

5 The unemployment rate over a period of a year is recorded and graphed. Two newspapers present the information using line graphs and their own newspaper heading as shown below.

LET'S TALK
The time on the y-axis is given as a decimal. How is time as a decimal converted to hours, minutes and seconds?

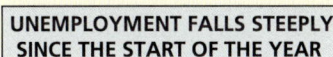

  a Are the newspapers showing the same information? Justify your answer.
  b Why might a newspaper decide to choose one graph rather than the other?

6 A book publisher wants to print information about the average number of pages a book has and the average number of lines each page has depending on the age group the book is aimed at.
You have been asked to present a short report to the publisher with this information.

a Carry out a detailed survey to analyse a number of books for different age groups and write a short report for the publisher with your findings.

b The publisher wants to use your findings to promote a new series of books aimed at teenagers. What title could you give to the report in order to support this?

> Ensure that your method for the data collection is clearly stated. The conclusion to the report should contain a **generalisation** about books for each different age group based on the samples you took.

7 A local newspaper decides to see how the price of cars changes depending on how old they are.
The reporter goes to the local garage and picks four cars at random, makes a note of their age and their sale price. He produces a scatter graph of the results and draws a line of best fit through the points as shown.

a i) Describe the correlation between sale price and car age seen on the graph.

ii) Is this correlation what you would expect to see? Justify your answer.

iii) State three other possible variables which may have influenced the correlation. Justify your choices.

b i) Carry out your own research into the relationship between a car's sale price and its age. State your method clearly and in particular how you minimised the effects of bias and spurious correlation.

ii) Plot a scatter graph of your results and state the correlation (if any) that it shows.

 Now you have completed Unit 12, you may like to try the Unit 12 online knowledge test if you are using the Boost eBook.

- Identify reflective symmetry in 3D shapes.
- Enlarge 2D shapes, from a centre of enlargement (outside, on or inside the shape) with a positive integer scale factor. Identify an enlargement, centre of enlargement and scale factor.
- Transform points and 2D shapes by combinations of reflections, translations and rotations.
- Identify and describe a transformation (reflections, translations, rotations and combinations of these) given an object and its image.
- Recognise and explain that after any combination of reflections, translations and rotations the image is congruent to the object.

Symmetry in three dimensions

So far all the cases of symmetry we have dealt with have been in two dimensions (i.e. the symmetry of 2D shapes).

However, 3D shapes also have symmetry properties. In this section we will look specifically at reflective symmetry of 3D shapes.

The reflective symmetry of a 2D shape is shown using a line of symmetry. The line of symmetry splits the shape into two **congruent** parts.

In a 3D shape, reflective symmetry is shown using a **plane of symmetry**.

A plane of symmetry divides a 3D shape into two congruent solid shapes.

The cuboid below has three planes of symmetry.

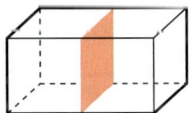 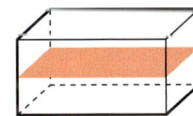 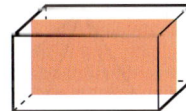

A 3D shape has reflective symmetry if it has one or more planes of symmetry.

> **KEY INFORMATION**
>
> Congruent shapes are identical in shape and size.

Exercise 13.1

1 For each of the following three-dimensional shapes, make two copies of the diagram shown. Then:

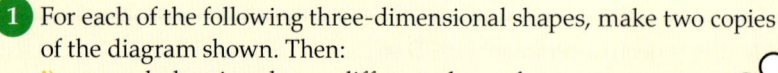

  i)   on each drawing show a different plane of symmetry

  ii) work out how many planes of symmetry the shape has in total.

> Use a ruler to draw your planes of symmetry.

a

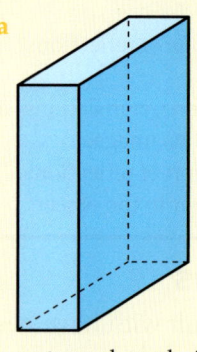

a rectangular cuboid

d

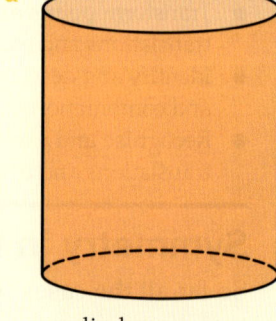

a cylinder

b

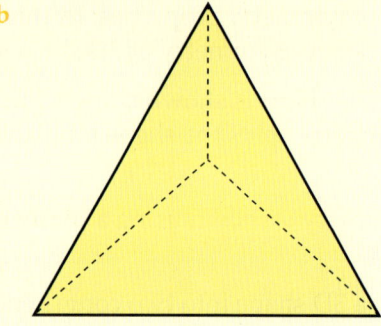

a triangular-based pyramid (tetrahedron)

e

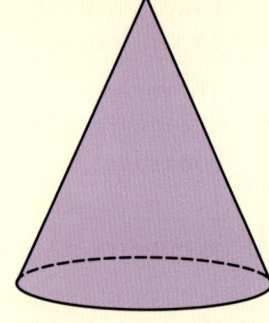

a cone

c

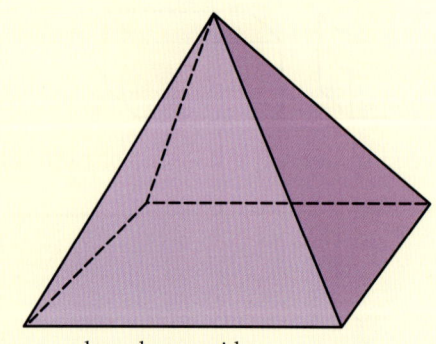

a square-based pyramid

f

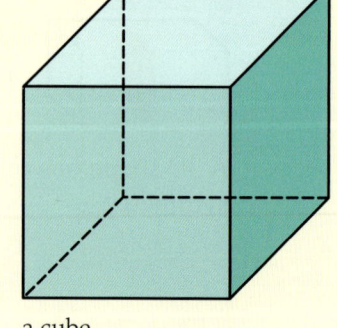

a cube

2 A cuboid is made from four cubes as shown.

  a **i)** How many planes of symmetry does the cuboid have?

     ii) Draw three of the planes of symmetry, each on a separate diagram.

  b Two extra cubes are to be added to the shape.

     i) Draw one position for the two cubes so that the new shape has just two planes of symmetry.
Draw one of the planes of symmetry on your diagram.

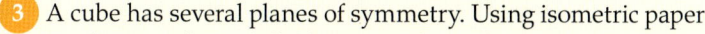

> You may find that the use of isometric dotted paper will help with drawing the cubes and cuboids.

     ii) Draw one position for the two cubes so that the new shape has just one plane of symmetry.
Draw the plane of symmetry on your diagram.

     iii) Draw one position for the two cubes so that the new shape has four planes of symmetry.
Draw one of the planes of symmetry on your diagram.

3 A cube has several planes of symmetry. Using isometric paper:

  a draw a cube and shade in one plane of symmetry

  b draw another cube and shade in another plane of symmetry

  c drawing a new cube each time and shading in a plane of symmetry, find the number of planes of symmetry a cube has.

4 This triangular prism has a constant cross-section in the shape of a right-angled isosceles triangle.

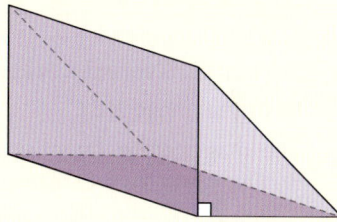

  a How many planes of symmetry does the shape have?

  b Two students are discussing the reflective symmetry of triangular prisms. They think that the type of triangle that makes the constant cross-section may affect the number of planes of symmetry of the prism itself.
Does the type of triangle affect the number of planes of symmetry?
Give **convincing** arguments for your answer.

Enlargement

You may recall from Stage 8 that enlargement is a type of transformation. When an object is enlarged, the image produced is mathematically **similar** but of a different size.

When describing an enlargement, two pieces of information need to be given. These are the position of the **centre of enlargement** and the **scale factor of enlargement**.

> **KEY INFORMATION**
> Although called enlargement, mathematically it is possible to produce an image that is smaller than the object, and it would still be classed as an enlargement. This section, though, will only deal with cases where the image is bigger than the object.

1   The grid below shows a triangle ABC.

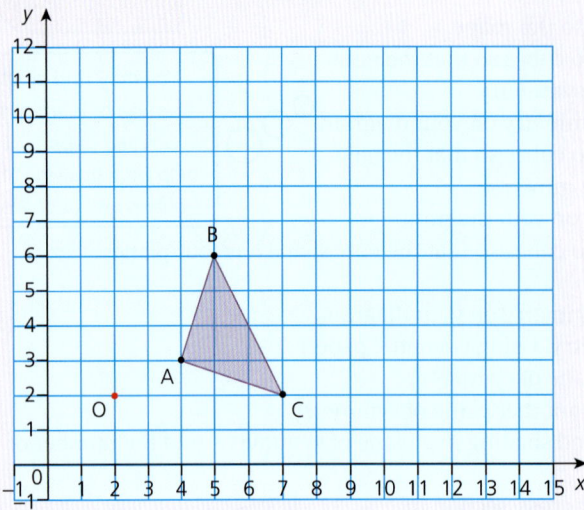

The triangle is enlarged by a scale factor of 2 from a centre of enlargement O with coordinates (2, 2).

Draw the image A′B′C′.

The image can be drawn as follows:

● Draw rays from O through each of the vertices A, B and C as shown.

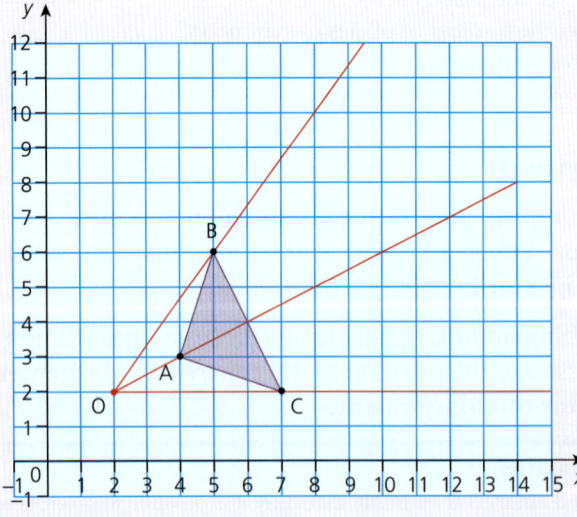

● With a scale factor of enlargement of 2, the distance from the centre of enlargement to the vertices of the shape is doubled. This can be calculated, for example, by using translation vectors.

Vector from O to $A = \begin{pmatrix} 2 \\ 1 \end{pmatrix}$ therefore vector from O to $A' = \begin{pmatrix} 4 \\ 2 \end{pmatrix}$

Vector from O to $B = \begin{pmatrix} 3 \\ 4 \end{pmatrix}$ therefore vector from O to $B' = \begin{pmatrix} 6 \\ 8 \end{pmatrix}$

Vector from O to $C = \begin{pmatrix} 5 \\ 0 \end{pmatrix}$ therefore vector from O to $C' = \begin{pmatrix} 10 \\ 0 \end{pmatrix}$

The enlargement is therefore:

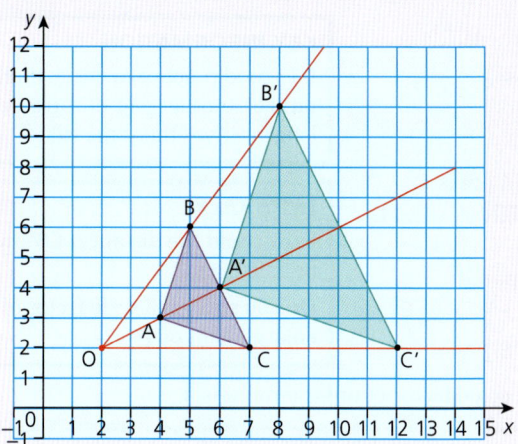

It is good practice to leave the rays used for the construction in the final diagram.

Notice how the angles remain unchanged during an enlargement.

Note that the ratio $\dfrac{A'B'}{AB} = \dfrac{A'C'}{AC} = \dfrac{B'C'}{BC} = 2$

LET'S TALK

What do these ratios mean? How do they relate to the enlargement diagram?

2   The image X'Y'Z' is an enlargement of the object XYZ.

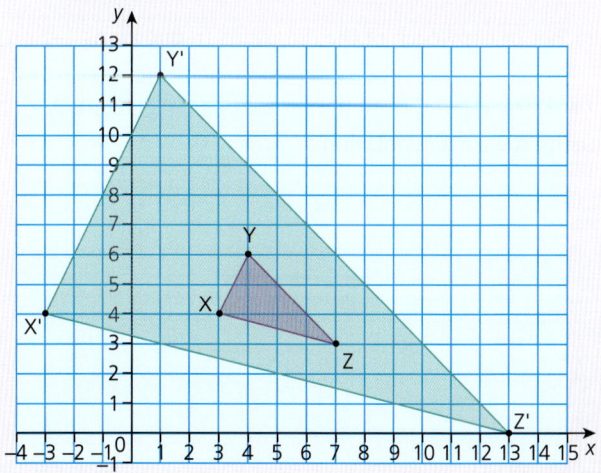

a Deduce the coordinates of the centre of enlargement.

This can be done by drawing rays through each pair of corresponding vertices, XX′, YY′ and ZZ′, extending them and seeing where they intersect. The point where they all intersect is the centre of enlargement O.

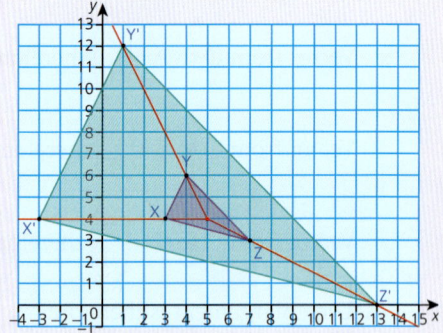

b Work out the scale factor of enlargement.

This can be done by looking at ratios:

$$\frac{OX'}{OX} = \frac{8}{2} = 4$$

Therefore the scale factor of enlargement is 4. However, not all lengths are as straightforward to calculate as the one above.

Translation vectors can also be compared:

$$O \text{ to } Y = \begin{pmatrix} -1 \\ 2 \end{pmatrix} \qquad O \text{ to } Y' = \begin{pmatrix} -4 \\ 8 \end{pmatrix}$$

It can be deduced from this that the scale factor of enlargement is 4.

KEY INFORMATION

It is good practice to check that the ratio of another pair of corresponding lengths gives the same value.

LET'S TALK

How could the distances OY and OY′ be calculated?

Do you still get a scale factor of enlargement of 4?

Exercise 13.2

Teachers are able to visit boost-learning.com where downloadable worksheets are available for the following questions.

Use the worksheets provided by your teacher or copy each of the diagrams in questions 1–2 on to squared paper.

Enlarge each of the objects by the given scale factor and from the centre of enlargement O.

Label each of the vertices of the image using the correct notation.

1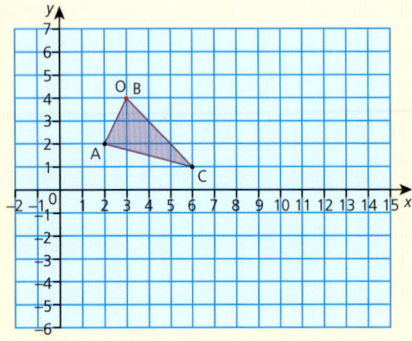

Scale factor of enlargement 3

2

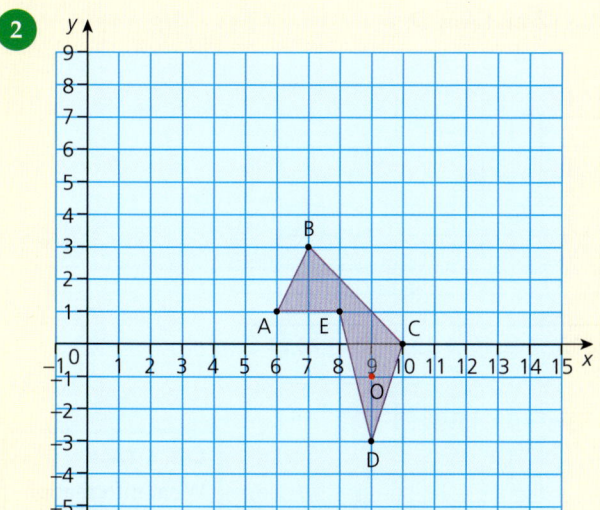

Scale factor of enlargement 2

3 The coordinates of the vertices of a quadrilateral PQRS are given as P(–4, –2), Q(–4, 2), R(2, 2) and S(4, 0). The shape is enlarged by a scale factor of 5 from a centre of enlargement O at (–2, –1). Calculate the coordinates of the image P'Q'R'S'.

4 Use the worksheet provided by your teacher or copy the object and image below.

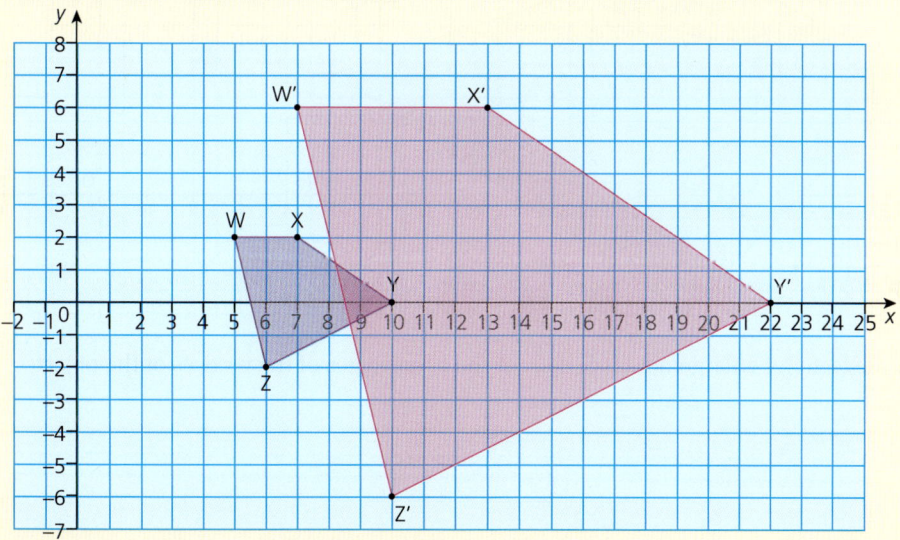

    a Explain whether the centre of enlargement will be to the left of the object WXYZ or to the right of the image W'X'Y'Z'.

    b Deduce, by construction, the position and coordinates of the centre of enlargement O.

    c Calculate the scale factor of enlargement.

5 The diagram below shows two shapes WXYZ and ABCD.

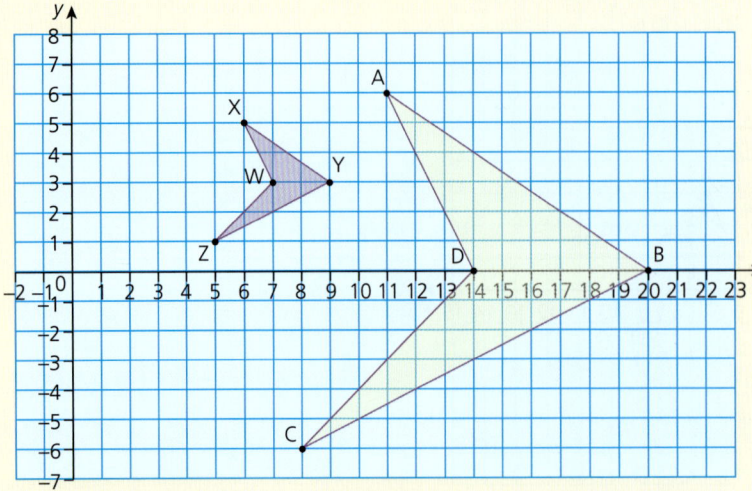

LET'S TALK

What information is needed here for it to count as a proof?

a   Prove numerically that ABCD is an enlargement of WXYZ.
b   Show, by construction, the position of the centre of enlargement.

6 The diagram below shows a centre of enlargement O at (12, 0) and an image W'X'Y'Z'.

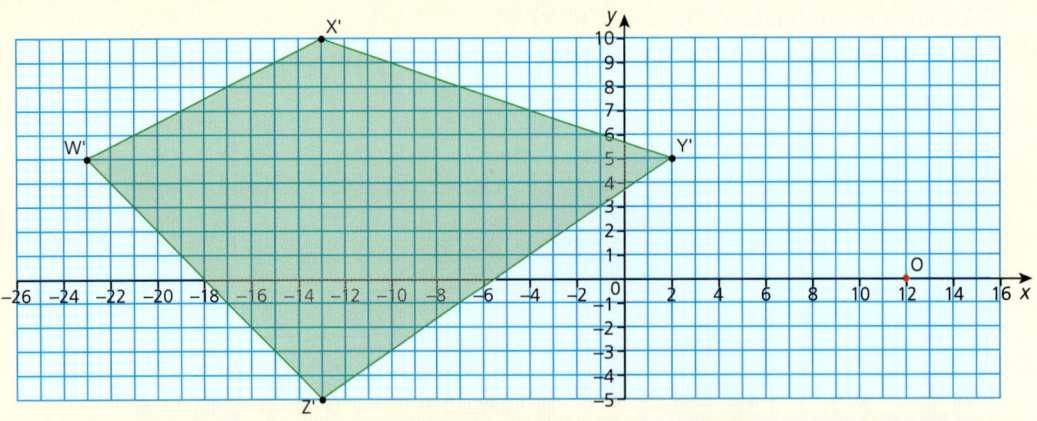

If the scale factor of enlargement is 5, deduce the coordinates of the vertices of the object WXYZ.

Combinations of transformations

You will now be aware of the individual transformations of reflection, rotation, translation and enlargement. It is also possible to apply multiple transformations to an object so that it undergoes more than one transformation to map it on to an image.

The types of transformation and the order in which they are carried out affect where the image appears and whether the object and image are congruent.

The trapezium X undergoes two transformations:

- a translation of $\begin{pmatrix} 5 \\ -2 \end{pmatrix}$
- a reflection in the line $y = -x$

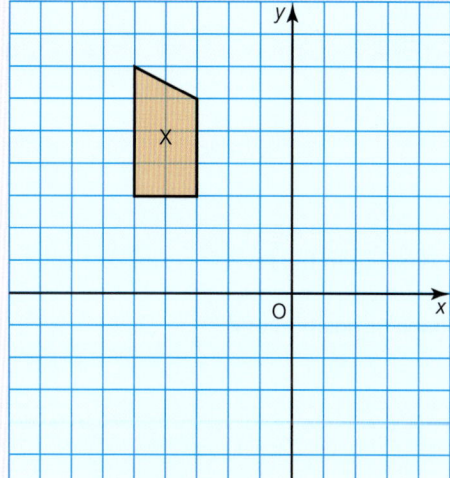

Draw the images, labelling the image after the first transformation Y and the image after the second transformation Z.

To draw the transformations, use the following steps.

- The translation has been described as a vector. The top number gives the horizontal movement and the bottom number gives the vertical movement. So image Y is located by translating each vertex of X by 5 units to the right and −2 units up (i.e. 2 units down).
- The reflection has been described by the equation of the mirror line on the coordinate grid. The straight line $y = -x$ is a downward sloping straight line through the origin. Draw this on the coordinate grid and reflect Y in it to locate the image Z.

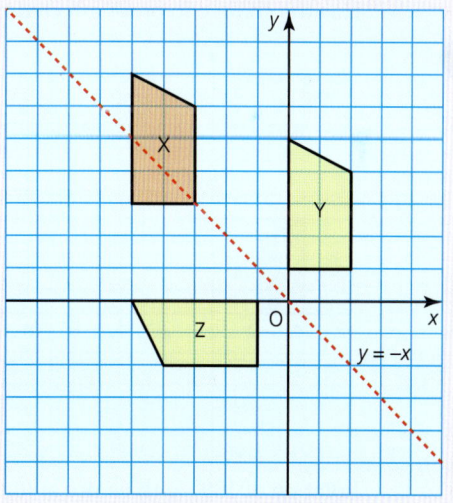

Exercise 13.3

Teachers are able to visit boost-learning.com where downloadable worksheets are available for the following questions.

In each of questions 1–4, the object X undergoes two transformations.

The first transformation maps X on to an image Y, the second maps Y on to an image Z.

Use the worksheets provided by your teacher or make two copies of each diagram on to squared paper and draw each of the images Y and Z, labelling them clearly.

1 a i)

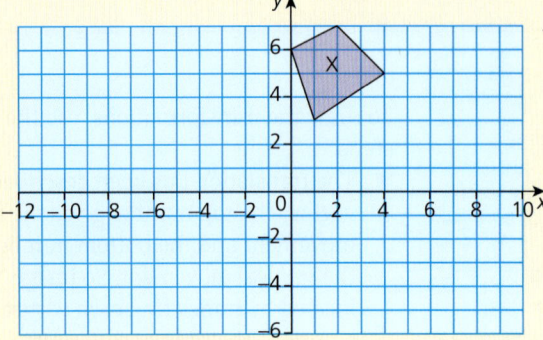

Reflection in the line $y = 3$

Translation of $\begin{pmatrix} -6 \\ -3 \end{pmatrix}$

  ii) Are X, Y and Z congruent?

b i) On your second grid repeat the transformations, but in the reverse order, i.e. the translation first followed by the reflection.

  ii) Has the change in order affected the final shape or position of Z?

2 a i)

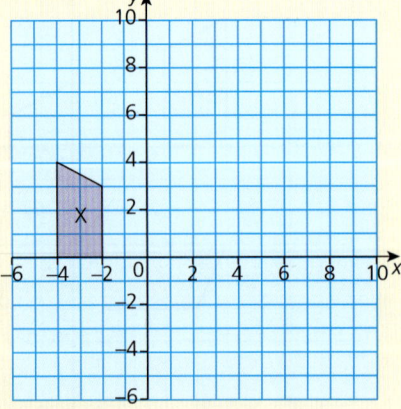

Rotation of 180° with centre at (0, 2)

Reflection in the line $y = x + 2$

  ii) Are X, Y and Z congruent?

b i) On your second grid repeat the transformations, but in the reverse order, i.e. the reflection first followed by the rotation.

   ii) Has the change in order affected the final shape or position of Z?

3 a i)

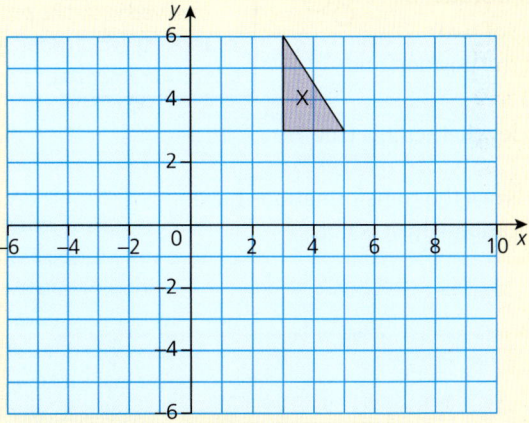

Enlargement with scale factor 2 and centre of enlargement at (2, 7)
Rotation of 90° clockwise about the origin

   ii) Are X, Y and Z congruent?

b i) On your second grid repeat the transformations, but in the reverse order.

   ii) Has the change in order affected the final shape or position of Z?

4 a i)

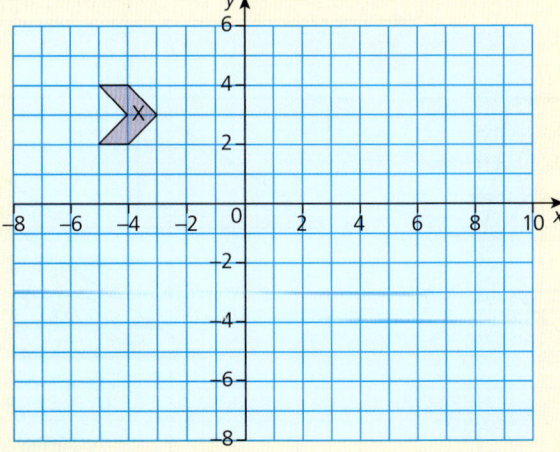

Reflection in the line $y = 0$
Enlargement with scale factor 3 and centre of enlargement at (−7, −5)

   ii) Are X, Y and Z congruent?

b i) On your second grid repeat the transformations, but in the reverse order.

   ii) Has the change in order affected the final shape or position of Z?

5 In questions 1–4, what type of transformation(s) led to the images not being congruent? Give a **convincing** reasoned argument why this is the case.

6 The following statement is made by a student:

'There are cases where the order of two transformations does not affect the position and shape of the final image.'

> This question can be treated as a mini investigation. There is potentially a lot of mathematics involved with the proof and reasoning, which does go beyond the requirements of the syllabus.

Investigate this statement and decide whether it is true or not.

If it is true, explain what **characteristics** the transformations have.

If it is not true, try and give a justification why it is not.

Exercise 13.4

In each of the following questions, the object X undergoes two transformations. The first transformation maps X on to the image Y, the second maps Y on to the image Z.

> For a reflection, give the equation of the mirror line.
> For a rotation, give the angle, the direction and the coordinates of the centre of rotation.
> For an enlargement, give the scale factor and the coordinates of the centre of enlargement.

Fully describe the two transformations in each case.

1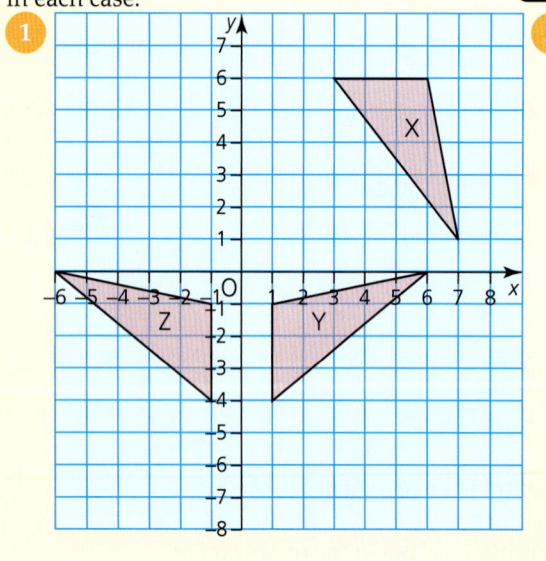

2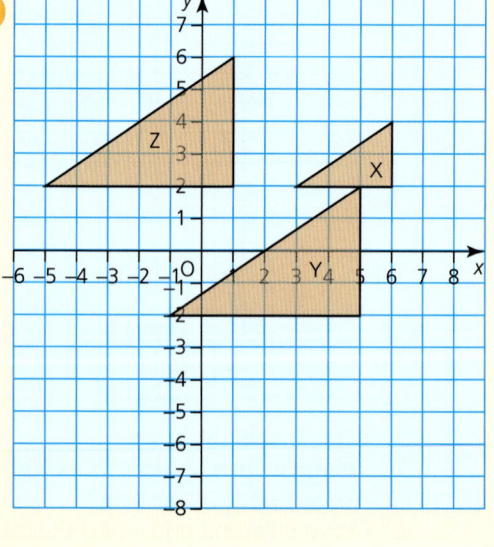

Now you have completed Unit 13, you may like to try the Unit 13 online knowledge test if you are using the Boost eBook.

Further fractions and decimals

- Deduce whether fractions will have recurring or terminating decimal equivalents.
- Estimate, add and subtract proper and improper fractions, and mixed numbers, using the order of operations.
- Estimate, multiply and divide fractions, interpret division as a multiplicative inverse, and cancel common factors before multiplying or dividing.
- Use knowledge of the laws of arithmetic, inverse operations, equivalence and order of operations (brackets and indices) to simplify calculations containing decimals and fractions.

Recurring and terminating decimals

Recap

You saw in Stage 8 that some fractions are equivalent to a **terminating decimal**, whilst others are equivalent to a **recurring decimal**.

Remember that a terminating decimal is one in which the digits after the decimal point stop, e.g. $\frac{1}{2} = 0.5$ or $\frac{1}{4} = 0.25$.

A recurring decimal, on the other hand, is one in which the digits after the decimal point continually repeat themselves (are **periodic**). This can be like $\frac{1}{3} = 0.\dot{3}$ which repeats the digit 3 straight after the decimal point, or $\frac{1}{45} = 0.0\dot{2}$ which repeats the digit 2 from the second decimal place onwards, or $\frac{1}{7} = 0.\dot{1}4285\dot{7}$ which repeats the group of digits 142857.

What this section will look at is how we can work out whether a fraction is equivalent to a recurring decimal or not.

One method is to see how it relates to a fraction which we already know produces a recurring decimal.

For example, is $\frac{2}{3}$ recurring?

We already know that $\frac{1}{3} = 0.\dot{3}$

$$\frac{2}{3} = 2 \times \frac{1}{3}$$
$$= 2 \times 0.\dot{3}$$
$$= 0.\dot{6}$$

> **LET'S TALK**
>
> What is the difference between a recurring decimal and an irrational number?
>
> Can you give some examples of irrational numbers?

> **LET'S TALK**
>
> Is the recurring decimal $0.\dot{9}$ equal to 1?
>
> Can you prove your answer?

Therefore $\frac{2}{3}$, as a multiple of $\frac{1}{3}$, is also recurring.

Is $\frac{1}{9}$ recurring?

We already know that $\frac{1}{3} = 0.\dot{3}$

$$\frac{1}{9} \times 3 = \frac{1}{3} \text{ therefore } \frac{1}{9} = \frac{1}{3} \div 3$$

$$= 0.\dot{3} \div 3$$

$$= 0.\dot{1}$$

Therefore $\frac{1}{3}$, as a multiple of $\frac{1}{9}$ shows that $\frac{1}{9}$ is also recurring.

However, simply being a multiple of a fraction that produces a recurring decimal is not itself sufficient to produce a recurring answer.

LET'S TALK

Why do multiples of recurring decimals not always produce recurring answers?

Find some exceptions. Can you **generalise**?

Worked example

a With a calculator show that $\frac{1}{6}$ is a recurring decimal.

Working out $1 \div 6$, the answer is $0.1\dot{6}$ so is recurring.

b Explain why $\frac{5}{6}$ is also recurring.

$$\frac{5}{6} = 5 \times \frac{1}{6}$$

$$= 5 \times 0.1\dot{6}$$

$$= 0.8\dot{3}$$

LET'S TALK

Your calculator may give the answer as 0.1666666667.

Why does it show a 7 at the end?

c Explain why $\frac{12}{6}$ is not equivalent to a recurring decimal.

$\frac{12}{6}$ can be simplified to 2.

2 is an integer and not a recurring decimal.

Parts (c) and (d) may help you understand why some multiples of fractions are equivalent to recurring decimals whilst others are not.

d Explain why $\frac{3}{6}$ is not equivalent to a recurring decimal.

$\frac{3}{6}$ can be simplified to $\frac{1}{2}$.

$\frac{1}{2}$ is equivalent to the terminating decimal 0.5 and is not a recurring decimal.

Exercise 14.1

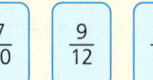

$\dfrac{7}{10}$ $\dfrac{9}{12}$ $\dfrac{11}{12}$ $\dfrac{1}{2}$ $\dfrac{4}{11}$ $\dfrac{1}{5}$ $\dfrac{3}{7}$ $\dfrac{5}{9}$ $\dfrac{3}{5}$ $\dfrac{7}{15}$ $\dfrac{6}{15}$ $\dfrac{9}{20}$

a Use a calculator to work out which of these fractions produce a terminating decimal and which produce a recurring decimal.
Classify them in a table similar to the one shown.

Terminating	Recurring

b Write a possible distinction between the fractions in the different columns of the table.

c i) Write another two fractions to go in each column of the table based on your answer to part (b).

ii) Use a calculator to check that your fractions are in the correct column of the table.

$\dfrac{4}{6}$ $\dfrac{10}{18}$ $\dfrac{14}{21}$ $\dfrac{14}{15}$ $\dfrac{5}{11}$ $\dfrac{12}{15}$

$\dfrac{7}{18}$ $\dfrac{8}{18}$ $\dfrac{3}{21}$ $\dfrac{8}{7}$ $\dfrac{9}{6}$ $\dfrac{15}{32}$

LET'S TALK

As a challenge, can you decide whether $\dfrac{15}{31}$ is a terminating or recurring decimal?
Can you prove your answer using division?

Does a calculator or spreadsheet help?

a Without using a calculator, decide which of these fractions produce a terminating decimal and which produce a recurring decimal.
Classify them in a table similar to the one shown.

Terminating	Recurring

b Check your groupings using a calculator.
Explain your method of classification.

3 For each of the statements below:
i) decide whether the statement is:
- always true
- sometimes true
- never true

LET'S TALK

Discuss your answers to these statements with another student and give a justification for your answer.

ii) if it is sometimes true, give one example of when it is true and one example of when it isn't true.

a A fraction with an odd number for the denominator produces a recurring decimal.

b A fraction in its lowest terms with an even number for the denominator produces a recurring decimal.

c A fraction with a prime number for the denominator produces a recurring decimal.

d A fraction with an even number for the denominator produces a terminating decimal.

e A fraction in its lowest terms with a multiple of a prime number for the denominator, except for 2 or 5, produces a recurring decimal.

f A recurring decimal will terminate eventually if carried on for long enough.

This is the complete rule for determining that a fraction will produce a recurring decimal.

From the work on the previous page, the following conclusion can be made:

A fraction in its lowest terms with a multiple of a prime number for the denominator, except for 2 or 5, will always produce a recurring decimal.

Calculations with fractions and decimals

The work on fractions and decimals covered before introduced you to the addition and subtraction of fractions written both in improper fraction form and in mixed number form.

This section will consolidate that and extend it further to include multiplication and division of these fractions, taking into account also the order of operations as stated using BIDMAS.

It will also look at methods of simplifying the calculations before carrying them out.

Addition and subtraction

Recap

The rectangles below are each divided into quarters. Seven sections in total are coloured.

 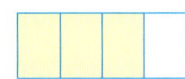

Seven quarters have been coloured, i.e. $\frac{7}{4}$ (an improper fraction as the numerator is larger than the denominator).

Similarly, it can be seen that one whole and three quarters have been coloured, i.e. $1\frac{3}{4}$ (a mixed number as it contains an integer and a proper fraction).

Therefore $\frac{7}{4} = 1\frac{3}{4}$

Converting this proper fraction to the mixed number form can be done as follows:

- See how many times the denominator goes into the numerator. In this case it is **1**, and this forms the integer part of the mixed number.
- Work out the remainder from the division. In this case it is **3**, and this forms the numerator of the proper fraction part of the mixed number.
- The improper fraction was written in quarters and this is still the case, so the 4 will be the denominator of the proper fraction part of the mixed number.

Therefore $\dfrac{7}{4} = 1\dfrac{3}{4}$

Converting this mixed number to improper fraction form can be done as follows:

- The integer 1 can be written in terms of quarters, i.e. $\dfrac{4}{4}$

- The mixed number can be converted to a sum, i.e. $\dfrac{4}{4} + \dfrac{3}{4}$

- The answer to the sum is $\dfrac{7}{4}$

Therefore $1\dfrac{3}{4} = \dfrac{7}{4}$

Worked examples

1 Work out the answer to the following calculation:

$$2\dfrac{3}{5} - \left(\dfrac{1}{4} + \dfrac{2}{3}\right)$$

LET'S TALK

How could this calculation be written without the brackets?

This can be worked out as follows:

- $2\dfrac{3}{5} - \left(\dfrac{3}{12} + \dfrac{8}{12}\right)$

  Due to BIDMAS the sum in the brackets is calculated first. First each fraction is written as an equivalent fraction with the same denominator.

- $2\dfrac{3}{5} - \dfrac{11}{12}$

  The sum in the brackets is worked out.

- $\dfrac{10}{5} + \dfrac{3}{5} - \dfrac{11}{12} = \dfrac{13}{5} - \dfrac{11}{12}$

  The mixed number is converted to an improper fraction.

- $\dfrac{13}{5} - \dfrac{11}{12} = \dfrac{156}{60} - \dfrac{55}{60}$

  To subtract the fractions, the denominators need to be the same, therefore both fractions are written as equivalent fractions with the same denominator.

- $\dfrac{156}{60} - \dfrac{55}{60} = \dfrac{101}{60}$

  Check that the answer cannot be simplified.

2 Work out the answer to the following calculation:

$$3\dfrac{2}{3} + 1\dfrac{1}{5} - \dfrac{18}{4}$$

LET'S TALK

Is there another way of working this out? Which method do you prefer? Why?

One way of working this out is as follows:

- $3\dfrac{2}{3} + 1\dfrac{1}{5} - 4\dfrac{1}{2}$

  Convert all the fractions to mixed numbers.

- $3\dfrac{20}{30} + 1\dfrac{6}{30} - 4\dfrac{15}{30}$

  Write each proper fraction part as an equivalent fraction with the same denominator.

- $4\dfrac{26}{30} - 4\dfrac{15}{30}$

  Start adding the terms.

- $\dfrac{11}{30}$

  Complete the calculation and check that the answer cannot be simplified further.

Exercise 14.2

In questions 1–6 work out the answer to the calculations, showing your method clearly and giving your answer as

a a mixed number   b an improper fraction.

1 $4\frac{1}{4} - 1\frac{1}{8}$

3 $3\frac{8}{9} - \left(\frac{3}{4} + \frac{5}{12}\right)$

5 $\frac{4}{3} + \left(\frac{2}{3} - \frac{3}{4}\right)$

2 $5\frac{2}{5} + \frac{10}{3}$

4 $6\frac{1}{5} - \left(1\frac{1}{3} + 3\frac{1}{2}\right)$

6 $2\frac{1}{6} - \left(\frac{14}{3} + 1\frac{3}{5}\right)$

7 A catering company has a stock of 15 blocks of butter.
They are asked to produce three different wedding cakes.
The first cake requires three whole blocks and two-thirds of another block.
The second cake requires five whole blocks and four-fifths of another block.
The third cake only requires two whole blocks.
  a i) Write a calculation using brackets to work out how much butter the company will have left after making all three cakes.
      ii) Write the same calculation without using brackets.
      iii) Which calculation will you find easier to work with? Justify your answer.
  b Calculate the amount of butter the company will have left.

Multiplication and division

In many ways multiplication and division involving fractions is easier than addition and subtraction as we do not have to worry about the denominators being the same.

There are also a few methods which can be used which could make the calculation more straightforward.

Worked examples

1 Calculate the answer to the following multiplication:
$\frac{9}{5} \times 1\frac{3}{7}$

This can be calculated as follows:

Method 1:

- $\frac{9}{5} \times \frac{10}{7}$ *Convert any mixed numbers to improper fractions.*

- $\frac{90}{35}$ *Multiply the numerators together and multiply the denominators together.*

- $\frac{18}{7}$ *Simplify the fraction if possible.*

It is not always easy to spot whether a fractional answer can be simplified. To make the calculation easier and to avoid having to do any simplifying at the end, it is usual to simplify fractions during the calculation itself. Looking again at the calculation above:

Method 2:

- $\dfrac{9}{5} \times 1\dfrac{3}{7}$

  Convert any mixed numbers to improper fractions as before in the mulitplication.

- $\dfrac{9}{5} \times \dfrac{10}{7}$

- $\dfrac{9}{1\!\!\!\;5} \times \dfrac{10^{\,2}}{7}$ *Simplify the fractions.*

- $\dfrac{9}{1} \times \dfrac{2}{7} = \dfrac{18}{7}$ *Multiply the fractions.*

> Notice how the answer is already given in its simplified form as all the simplifying was carried out earlier in the calculation.

LET'S TALK

Why is it mathematically correct to simplify a multiplication like this

$$\dfrac{9}{1\!\!\!\;5} \times \dfrac{10^{\,2}}{7}$$

but not an addition like this $\dfrac{9}{1\!\!\!\;5} + \dfrac{10^{\,2}}{7}$?

2 Calculate the answer to the following division:

$$1\dfrac{2}{9} \div 7\dfrac{1}{3}$$

This can be calculated as follows:

- $\dfrac{11}{9} \div \dfrac{22}{3}$ *Convert the mixed numbers to improper fractions.*

- $\dfrac{11}{9} \times \dfrac{3}{22}$ *Rewrite the division as a multiplication. Remember that a division by a fraction is the same as multiplying by its reciprocal.*

- $\dfrac{11^{\,1}}{3\,9} \times \dfrac{3^{\,1}}{2\,22}$ *Simplify the fractions in the multiplication.*

- $\dfrac{1}{3} \times \dfrac{1}{2} = \dfrac{1}{6}$ *Multiply the simplified fractions together and double check that the answer cannot be simplified further.*

3 Work out the answer to the following calculation:

$$4\dfrac{1}{5} - 2\dfrac{1}{4} \div 1\dfrac{1}{2}$$

This can be calculated as follows:

- $\dfrac{21}{5} - \dfrac{9}{4} \div \dfrac{3}{2}$ *Convert the mixed numbers to improper fractions.*

- $\dfrac{21}{5} - \dfrac{9}{4} \div \dfrac{3}{2}$ *Identify what must be worked out first according to BIDMAS.*

- $\dfrac{21}{5} - \dfrac{9}{4} \times \dfrac{2}{3}$ *Rewrite the division as a multiplication.*

- $\dfrac{21}{5} - \dfrac{9^{\,3}}{2\,4} \times \dfrac{2^{\,1}}{1\,3}$ *Simplify the calculation by identifying the highest common factors.*

- $\dfrac{21}{5} - \dfrac{3}{2} \times \dfrac{1}{1}$ *Rewrite the calculation in simplified form.*

- $\dfrac{21}{5} - \dfrac{3}{2}$ *Work out the multiplication part of the calculation.*

- $\dfrac{42}{10} - \dfrac{15}{10}$ *Rewrite using equivalent fractions with the same denominator.*

- $\dfrac{27}{10}$ *Work out the final answer and double check that it cannot be simplified further.*

4   Work out the answer to the following calculation involving a mixture of fractions and decimals:

$$(3.2)^2 + 4 \times 3\dfrac{1}{2}$$

This can be calculated as follows:

- $\left(3\dfrac{1}{5}\right)^2 + 4 \times 3\dfrac{1}{2}$ *Convert decimals to fractions.*

- $\left(\dfrac{16}{5}\right)^2 + 4 \times \dfrac{7}{2}$ *Convert mixed numbers to improper fractions.*

- $\dfrac{256}{25} + 4 \times \dfrac{7}{2}$ *The indices part is worked out first due to BIDMAS.*

- $\dfrac{256}{25} + 14$ *The multiplication is worked out next due to BIDMAS.*

- $10\dfrac{6}{25} + 14$ *Due to the numbers involved, it is probably easier to work with mixed numbers rather than improper fractions.*

- $24\dfrac{6}{25}$ *Work out the final answer and double check it cannot be simplified further.*

Exercise 14.3

For questions 1 to 5 calculate the following, showing your method clearly and giving your answer in its simplest form.

1 $\dfrac{2}{7} \times 3\dfrac{1}{2}$

2 $\dfrac{5}{3} \times 1\dfrac{1}{4}$

3 $4\dfrac{1}{8} \times 2\dfrac{2}{11}$

4 $1\dfrac{1}{6} \div \dfrac{5}{18}$

5 $7\dfrac{5}{8} \div \dfrac{5}{16}$

 6   Imperial weights use pounds (lbs) and ounces (oz).
16 oz are equivalent to 1 lb.
- a   A large bag of sweets has a mass of 3 lb and 2 oz.
  Write this in pounds using fractions.
- b   The bag of sweets is to be shared equally amongst 5 friends.
  Work out how much each receives, giving your answer in
     i)   pounds   ii)   ounces.

For questions 7 to 11 calculate the following, showing your method clearly and giving your answer in its simplest form.

7 $1\frac{1}{2} - \left(\frac{3}{8} + 1\frac{3}{4}\right)$

10 $\left(5\frac{4}{9} \div \frac{21}{6}\right)^2 \times \frac{9}{14}$

8 $1\frac{7}{10} - \frac{4}{5} \times 1\frac{3}{8}$

11 $1 - \left(\frac{8}{15}\right)^2 \div \frac{16}{25}$

9 $8\frac{9}{16} + \frac{3}{4} \div 1\frac{5}{7}$

 12 Imperial measurements for distance include yards, feet and inches.

There are 12 inches in 1 foot and 3 feet in 1 yard.

Two tortoises A and B set off at the same time and crawl in the same direction along the same straight line. Tortoise A starts 1 foot and 6 inches ahead of tortoise B.

After 2 minutes tortoise B has crawled 4 feet and 8 inches, whilst in the same time tortoise A has travelled twice that distance.

a   What is the total distance travelled by the two tortoises after 2 minutes? Give your answer in
   i) yards   **ii)** feet   **iii)** inches.

b   Assuming the tortoises move at a constant speed, calculate the distance between them after 5 minutes. Give your answer in
   i) yards   **ii)** feet   **iii)** inches.

For questions 13 to 15 calculate the following, showing your method clearly and giving your answer in its simplest form.

13 $5.12 \div \left(\frac{2}{5}\right)^2$

14 $\left(2.1 - 1\frac{1}{5}\right)^2 - 1\frac{3}{5} \times \frac{1}{2}$

15 $30 - \frac{7.25}{8\frac{1}{2} - \frac{9}{4}} \times 5^2$

 16 Imperial weights include stones, pounds and ounces.

● 1 stone is equal to 14 pounds.

● 1 pound is equal to 16 ounces.

A container holds 2 stones, 8 pounds and 4 ounces of flour.

The equivalent of 2.5 containers are supplied to a local bakery. However, on the way a total of 45 ounces of flour is spilt.

a   Express the mass of one container in
   i) pounds   **ii)** stones.

b   What is the total mass of flour delivered to the shop?
   Give your answer in
   i) pounds   **ii)** ounces.

Now you have completed Unit 14, you may like to try the Unit 14 online knowledge test if you are using the Boost eBook.

Manipulating algebraic expressions

- Understand how to manipulate algebraic expressions including:
  - expanding the product of two algebraic expressions
  - applying the laws of indices
  - simplifying algebraic fractions.

Expanding the product of two algebraic expressions

In Stage 8 you were introduced to the **distributive law** of multiplication, where the number outside a bracket multiplies each of the terms inside it.

It was applied in the context of expanding the expression $2a(5a+b)$.

Visually this can be represented as finding the area of a rectangle with dimensions of $5a+b$ and $2a$, the side of length $5a+b$ being divided into $5a$ and b as shown:

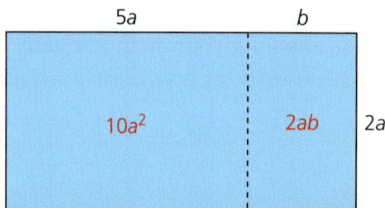

The area of each section is calculated:

$$2a \times 5a = 10a^2$$

$$2a \times b = 2ab$$

Therefore, the total area is $10a^2 + 2ab$.

But the area of a rectangle is the multiplication of the length by the width, so

$$2a(5a+b) = 10a^2 + 2ab$$

The expression can be given in expanded form by multiplying out the brackets. Using the distributive law this can be seen as:

$$2a(5a+b) = 10a^2 + 2ab$$

> Multiplying $2a \times 5a$ can be considered as $2 \times 5 \times a \times a$.

A similar method can be used for the product of more complicated expressions.

Worked example

Write an expression for the area of this rectangle.

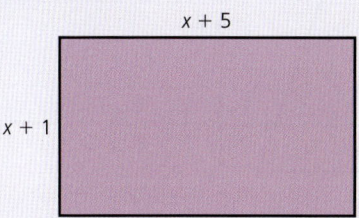

The rectangle can be split into four like this:

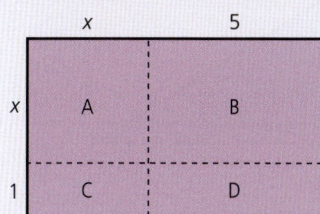

- The area of A is $x \times x = x^2$.
- The area of B is $5 \times x = 5x$.

- The area of C is $1 \times x = x$.
- The area of D is $1 \times 5 = 5$.

The total area is given by the expression

$$x^2 + 5x + x + 5 = x^2 + 6x + 5$$

The area of the rectangle can also be expressed using brackets like this:

$$(x+1)(x+5)$$

Therefore

$$(x+1)(x+5) = x^2 + 6x + 5$$

> **KEY INFORMATION**
>
> The expression written in the form $(x+1)(x+5)$ is in **factorised form**, whilst if written as $x^2 + 6x + 5$ it is called **expanded form**.

To expand brackets, you multiply all the terms in one set of brackets by the terms in the other set of brackets.

$$(x+1)(x+5) = x^2 + 5x + x + 5$$
$$= x^2 + 6x + 5$$

Exercise 15.1

For each shape in questions 1–7:
a   write an expression for the area, using brackets
b   expand your expression in part (a) by multiplying out the brackets.

1
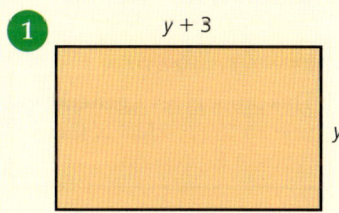

2
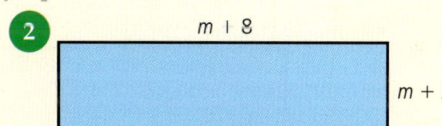

3

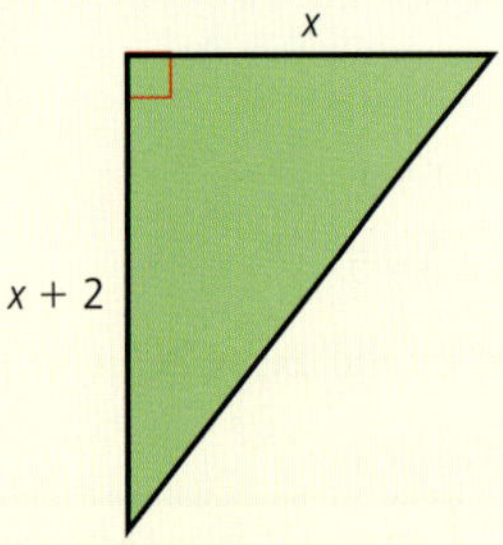

x

x + 2

6

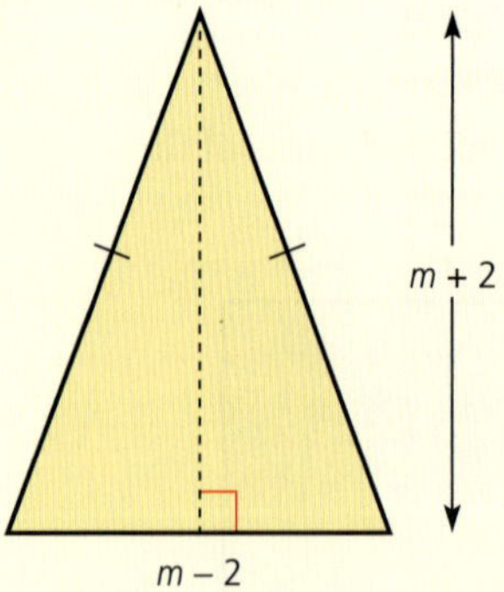

m + 2

m − 2

4

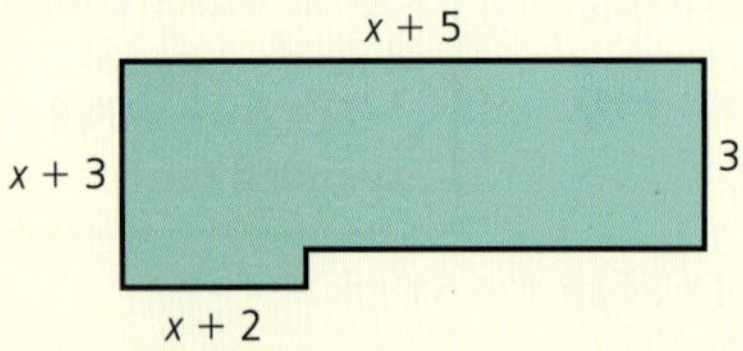

x + 5

x + 3

3

x + 2

7

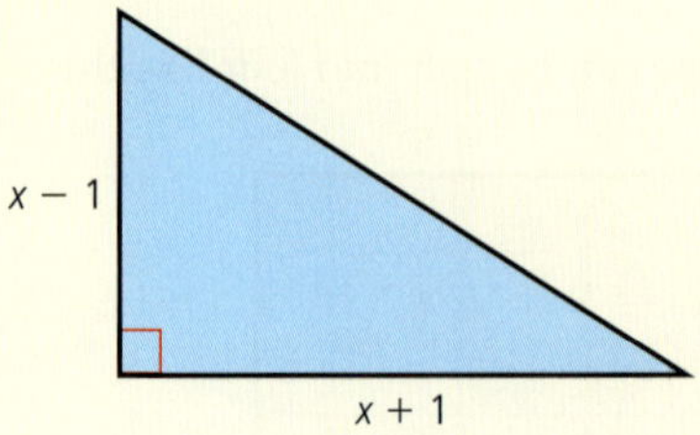

x − 1

x + 1

5

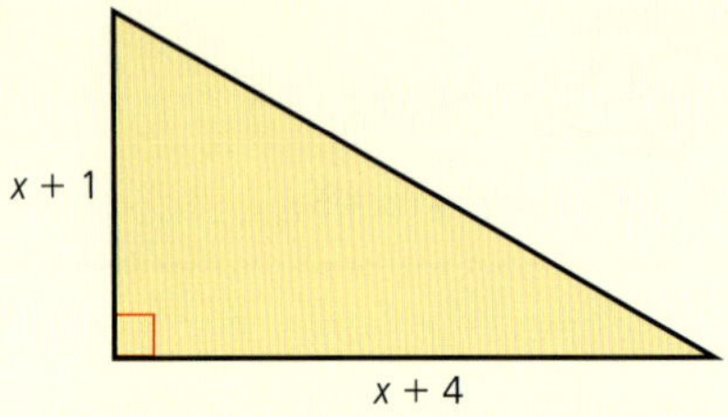

x + 1

x + 4

By considering the product of two algebraic expressions in general terms, it is possible to spot some patterns.

Consider the product of the two expressions $(x+a)$ and $(x+b)$.

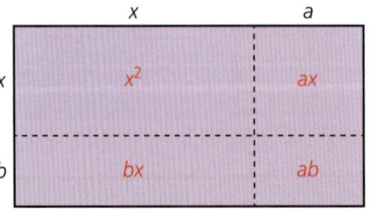

	x	a
x	x^2	ax
b	bx	ab

KEY INFORMATION

This proof gives a **general** rule for the product of these types of expression.

KEY INFORMATION

$ax+bx$ can be factorised to give $(a+b)x$.

From the areas, it can be deduced that $(x+a)(x+b)=x^2+ax+bx+ab$.

Therefore $(x+a)(x+b)=x^2+(a+b)x+ab$.

Looking at the expansion, it can be deduced that the sum $a+b$ becomes the **coefficient** of the x term, whilst the product ab is the constant.

Worked examples

1. By inspection, expand the product $(x+5)(x+7)$.

$$(x+5)(x+7) = x^2 + (5+7)x + (5 \times 7)$$
$$= x^2 + 12x + 35$$

2. By inspection, expand the product $(x-2)(x+6)$.

$$(x-2)(x+6) = x^2 + (-2+6)x + (-2 \times 6)$$
$$= x^2 + 4x - 12$$

> **KEY INFORMATION**
> 'By inspection' means to do it simply by looking at it, rather than carrying out the calculation.

There are some other special cases which are good to be aware of.

Worked examples

1. Expand $(x-3)(x+3)$.

$$(x-3)(x+3) = x^2 + 3x - 3x - 9$$
$$= x^2 - 9$$

> Notice that the solution consists of x^2 with 3^2 being subtracted from it.

Notice that the x term is not present in the expansion. This is because the numbers in the brackets are the same except for the change of sign, i.e. +3 and −3. On expansion they form the terms +3x and −3x, which cancel out.

This can be seen in more **general** terms with the expansion of $(x-a)(x+a)$:

$$(x-a)(x+a) = x^2 - ax + ax - a^2$$
$$= x^2 - a^2$$

> The solution consists of x^2 with a^2 being subtracted from it.

As this is a square term being subtracted from another square term, this is known as the **difference of two squares**.

2. Expand $(x+5)^2$.

This is the same as working out $(x+5)(x+5)$ so by inspection the expansion is:

$$(x+5)^2 = x^2 + 10x + 25$$

In **general**, this is $(x+a)^2 = x^2 + 2ax + a^2$

This expansion is known as a **perfect square**.

Exercise 15.2

1 Expand each of the following, by multiplication or by inspection. Give your answer in its simplest form.

a $(x+4)(x+6)$ e $(m-4)(m-7)$

b $(x+8)(x+3)$ f $(a-4)^2$

c $(y+1)(y-3)$ g $(2-x)^2$

d $(p-9)(p+9)$

2 The expanded form of an expression is given as $x^2 - 12x + 32$

a Identify which of the following expressions it is equal to.

   i) $(x+4)(x+8)$ iv) $(x-6)(x-2)$

   ii) $(x-7)(x-5)$ v) $(x-4)(x-8)$

   iii) $(x-9)(x-3)$ vi) $(x+2)(x+16)$

b Without multiplying them out, give a justification for why the other expressions are incorrect.

3 An expression for the area of each rectangle and the length of one of its sides is given.

Work out an expression for the width of the rectangle.

a

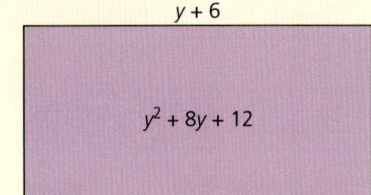

$y + 6$

$y^2 + 8y + 12$

b

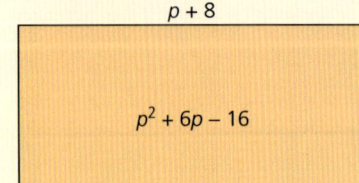

$p + 8$

$p^2 + 6p - 16$

4 An expression for the area of each rectangle and the length of one of its sides is given.

i) If it is possible, work out an expression for the width of the rectangle.

ii) If it is not possible, give a reasoned explanation why.

a

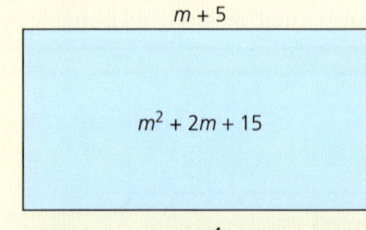

$m + 5$

$m^2 + 2m + 15$

b

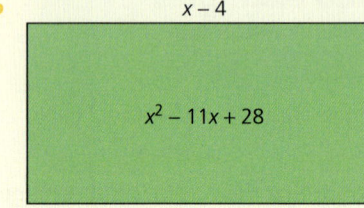

$x - 4$

$x^2 - 11x + 28$

5 a Show that $(x+a)^2 = x^2 + 2ax + a^2$.

b Use the expansion in part (a) to work out the following calculations.

   i) 1.5^2 ii) 34^2

c Show that $(x+a)(x-a) = x^2 - a^2$.

d Use the expansion in part (c) to work out 18×22

6 A rectangle has dimensions $(x+8)$ and $(x+6)$.

A rectangle A is cut from the larger rectangle and the shaded area left is given by the expression $6x + 32$.

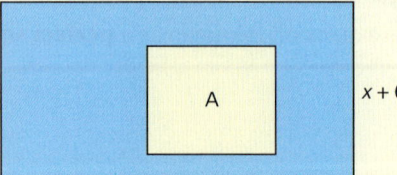

$x + 8$

A

$x + 6$

Work out whether the rectangle A is a square. Justify your answer.

Simplifying algebraic expressions

Recap

In Unit 1 of this book you looked at some of the laws of indices and what they meant. A summary of them is given below.

- $a^m \times a^n = a^{m+n}$ e.g. $3^5 \times 3^4 = 3^{5+4} = 3^9$
- $a^m \div a^n = a^{m-n}$ e.g. $3^8 \div 3^5 = 3^{8-5} = 3^3$
- $\left(a^m\right)^n = a^{mn}$ e.g. $\left(3^4\right)^2 = 3^{4\times2} = 3^8$
- $a^0 = 1$ e.g. $3^0 = 1$
- $a^{-m} = \dfrac{1}{a^m}$ e.g. $3^{-5} = \dfrac{1}{3^5}$

These rules enable some more complex expressions to be simplified.

Worked example

Simplify the following expression: $\dfrac{3n^7 \times 4n}{6n^5}$

As multiplication is commutative, the numerator can be simplified by writing it as $3 \times 4 \times n^7 \times n = 12 \times n^{(7+1)}$

KEY INFORMATION

'Commutative' means it can be done in any order and the answer will be the same.

Therefore, the original expression can be simplified to $\dfrac{12n^8}{6n^5}$

To simplify this further, it can be visualised as the product of two fractions as $\dfrac{12}{6} \times \dfrac{n^8}{n^5}$

which simplifies to $2 \times n^3$

Therefore, $\dfrac{3n^7 \times 4n}{6n^5} = 2n^3$

Exercise 15.3

1 Using the laws of indices, simplify the following.

 a $2a^0 \div a^7$

 b $4b^2 \times b^3 \div 2b^4$

 c $c^4 \times 6c^7 \div 2c^6$

 d $\left(p^2\right)^3 \div \left(p^3\right)^2$

 e $\left(2p^2\right)^3 \div \left(2p^3\right)^2$

2 Each of the diagrams below shows the length of two lines.

In each case write an expression for
 i) the total length of both lines
 ii) the difference in length between the two lines.

 a $3x^2 + y$ ____ $x + y$ ____

 b $2n^3 + 3m$ ____ $4m + n^3 + n$ ____

 c $3n^3 - 2m$ ____ $4m - n^3 + n$ ____

LET'S TALK

Is it possible to deduce which line is longer simply from the expressions for the lengths?

3 Simplify the following.

a $\left(m^3 \times 3m^5\right) \div \left(m^2 \times m^4\right)$

c $p^2 \times p^{-3} \times p^8 \div p^5 \times p^{-4}$

e $\left(b^{-4}\right)^2 \times \left(b^2\right)^4$

b $n^5 \times n^7 \times n \div n^6 \times n^{-2}$

d $\left(\dfrac{1}{a^3}\right)^2 \div a^2$

f $\left(c^2\right)^3 \times \left(\dfrac{1}{c^3}\right)^2 \div \left(c^5\right)^{-2}$

4 Write an expression for the area of each of the shapes below:

i) using brackets

ii) in simplified expanded form.

a

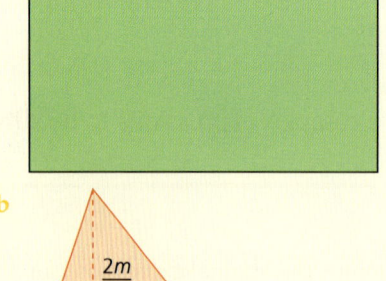

pq

p + q

c

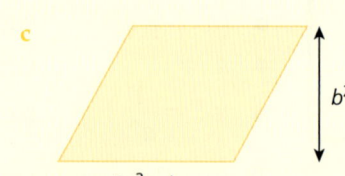

b^2

$3a^2 + b$

b

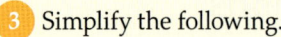

$\dfrac{2m}{n}$

$6m^2n^2$

d

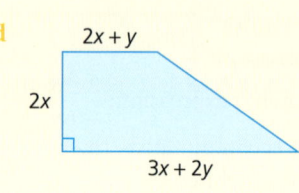

2x + y

2x

3x + 2y

You saw earlier that the factorised form of $(x+a)(x+b)$ could be expanded to give $x^2 + (a+b)x + ab$. Being able to factorise expressions often helps simplify calculations and solutions.

Worked examples

1 Simplify the expression $\dfrac{8x+20}{4}$

Method 1:

It is often easier to factorise an expression before simplifying it.

The numerator $8x + 20 = 4(2x+5)$

Therefore, $\dfrac{8x+20}{4} = \dfrac{4(2x+5)}{4}$

As both numerator and denominator are clearly now multiples of 4, these can be cancelled:

$\dfrac{\cancel{4}(2x+5)}{\cancel{4}} = 2x+5$

Method 2:

Another method to explain why the 4's cancel is shown below.

We have seen above that $\dfrac{8x+20}{4} = \dfrac{4(2x+5)}{4}$.

$\dfrac{4(2x+5)}{4}$ can also be expressed as a

product of fractions as $\dfrac{4}{4} \times \dfrac{2x+5}{1}$

As $\dfrac{4}{4} = 1$ and $\dfrac{2x+5}{1} = 2x+5$ we can

conclude that $\dfrac{4(2x+5)}{4} = 2x+5$

2 a Show that $(x-2)(x+6)=x^2+4x-12$

Multiplying each of the terms produces

$$(x-2)(x+6)=x^2-2x+6x-12$$
$$=x^2+4x-12$$

b Simplify the expression $\dfrac{x^2+4x-12}{x-2}$

Method 1:

It has already been shown that
$$x^2+4x-12=(x-2)(x+6)$$

The fraction can be therefore be written as $\dfrac{(x-2)(x+6)}{(x-2)}$

Both numerator and denominator are multiples of $(x-2)$ so these can be cancelled:

$$\dfrac{\cancel{(x-2)}(x+6)}{\cancel{(x-2)}}=x+6$$

Method 2:

The fraction $\dfrac{(x-2)(x+6)}{(x-2)}$ can be written as a product of the fractions as

$$\dfrac{(x-2)}{(x-2)}\times\dfrac{(x+6)}{1}$$

As $\dfrac{(x-2)}{(x-2)}=1$ and $\dfrac{(x+6)}{1}=x+6$, we can conclude that $\dfrac{(x-2)(x+6)}{(x-2)}=x+6$.

3 Simplify the expression $\dfrac{3n^2+2n}{2n}$

Method 1:

Factorising the numerator gives
$$\dfrac{3n^2+2n}{2n}=\dfrac{n(3n+2)}{2n}$$

Both numerator and denominator are multiples of n, so these can be cancelled to give

$$\dfrac{\cancel{n}(3n+2)}{2\cancel{n}}$$

Therefore, $\dfrac{3n^2+n}{2n}=\dfrac{3n+2}{2}$

Method 2:

The factorised expression $\dfrac{n(3n+2)}{2n}$ can be written as a product of two fractions as

$$\dfrac{n}{n}\times\dfrac{3n+2}{2}$$

$\dfrac{n}{n}=1$, whilst $\dfrac{3n+2}{2}$ cannot be simplified further.

Therefore $\dfrac{3n^2+2n}{2n}=\dfrac{3n+2}{2}$

Note that a common error is to try and cancel the answer down further.

The 2's cannot be cancelled like this:

$$\dfrac{3n+\cancel{2}}{\cancel{2}}$$

This is because they are not a common multiple of both the numerator and denominator.

However, there is another mathematically correct way of writing the answer.

$\dfrac{3n+2}{2}$ can be written as a sum of the two fractions $\dfrac{3n}{2}+\dfrac{2}{2}$.

As $\dfrac{2}{2}=1$, this can be simplified to $\dfrac{3n}{2}+1$

Therefore, $\dfrac{3n^2+2n}{2n}=\dfrac{3n+2}{2}=\dfrac{3n}{2}+1$

Exercise 15.4

1 Simplify the following expressions.

a $\dfrac{4x^2}{x}$ b $\dfrac{6y^3}{3y}$ c $\dfrac{2m^2n}{mn}$ d $\dfrac{5p^3q^2}{p^2q^3}$

2 The area and length of one side of this rectangle are given.
  a Write the calculation needed to work out the length marked '?'.
  b Simplify the calculation to work out the expression for '?'.

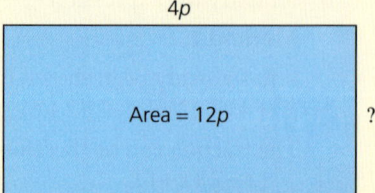

4p

Area = 12p   ?

3 Four rectangles are arranged as shown.
The total area of the four rectangles is given by the expression $16x^2 + 8x$.
If the width of one rectangle is $2x$, calculate the length of one of the rectangles.

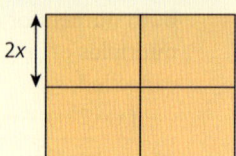

$2x$

4 i) Simplify, if possible, each of the following expressions.
  ii) If it cannot be simplified, give a reasoned answer why.

a $\dfrac{6+2x}{2}$ c $\dfrac{5x+6}{6}$

b $\dfrac{15m^2+5m}{10m}$ d $\dfrac{2(x+1)(x+4)}{2(x+4)}$

5 The tile pattern consists of a square and four congruent right-angled triangles.
The total area of the four triangles is given by the expression $4x^2 + 12x$.
  a Write an expression for the area of the square.
  b Write an expression for the total area of the shape, giving your answer in simplified form.

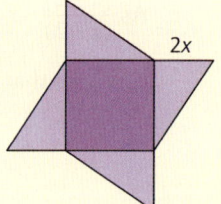

$2x$

6 a Show that $(x-6)^2 = x^2 - 12x + 36$.

  b Simplify the expression $\dfrac{x^2 - 12x + 36}{2x - 12}$

7 Simplify the following expressions.

a $\dfrac{x^2 - 49}{x+7}$ b $\dfrac{t^2 - 144}{2t - 24}$

Now you have completed Unit 15, you may like to try the Unit 15 online knowledge test if you are using the Boost eBook.

16 Combined events

- Identify when successive and combined events are independent and when they are not.
- Understand how to find the theoretical probabilities of combined events.

Independent events

When does the outcome of one event affect the outcome of another event?

When does the outcome of one event not affect the outcome of another event?

These are the questions we will be looking at in this section.

If the outcome of one event does not affect the outcome of another event, then they are said to be **independent events**.

Consider two normal, unbiased dice being rolled.

Does the result for one dice affect the result for the other?

Clearly what you get on one dice will not affect the result for the second dice.

Therefore, the results for the two dice are said to be independent events.

In a raffle, the winner can choose from one of three prizes.

The second winner can therefore choose from the two remaining prizes.

The third winner chooses the prize that is left.

LET'S TALK
When might raffle prizes be independent events?

These events are not independent. The third winner can only choose the prize that has been left by the other two. The second winner cannot pick the prize that the winner has chosen. Therefore the choice of one, in this case, is dependent on the choices of others.

Knowing whether events are independent or not is important for working out the probability of events happening.

Worked example

A bag contains ten sweets. Six are mint flavoured and four are lemon flavoured.

a Calculate the probability that the first sweet I choose is mint flavoured.

$$P(\text{mint}) = \frac{6}{10}$$

b If the first sweet I eat is mint flavoured, what is the probability that the second sweet I choose is mint flavoured?

There are now only nine sweets left in the bag, of which five are mint flavoured.

Therefore, $P(\text{mint}) = \frac{5}{9}$

c If the first sweet I eat is lemon flavoured, what is the probability that the second sweet I choose is mint flavoured?

There are now only nine sweets left in the bag, of which six are mint flavoured.

Therefore, $P(\text{mint}) = \frac{6}{9}$

d Is the picking of the second sweet independent of the first one? Justify your answer.

No, they are not independent. The probability of picking a mint sweet the second time changes depending on what was picked the first time.

The above example can be visualised using a tree diagram.

The outcomes are worked out by following the branches.

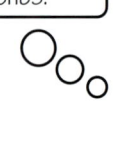

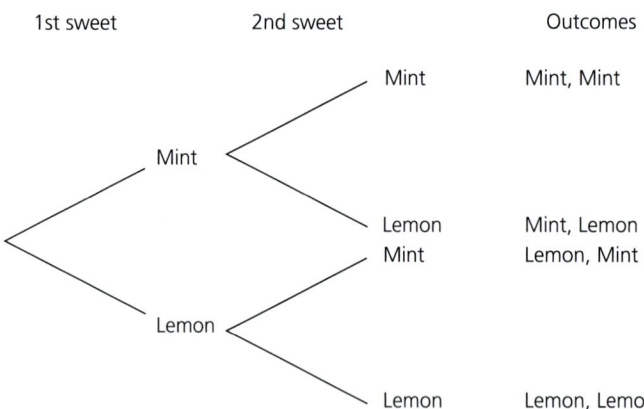

From the tree diagram it can be deduced that there are four possible outcomes for picking two sweets.

The diagram can be adapted further to include the probabilities as shown:

LET'S TALK

How do we use the diagram to calculate the probability of choosing two mint sweets?

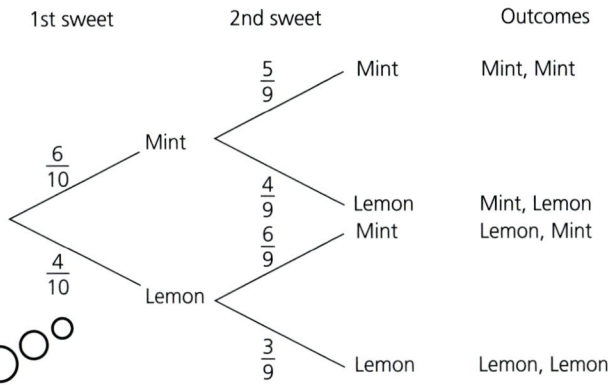

Notice how the sum of the joined branches at each stage is 1,

i.e. $\dfrac{6}{10}+\dfrac{4}{10}=1$

and $\dfrac{5}{9}+\dfrac{4}{9}=1$

and $\dfrac{6}{9}+\dfrac{3}{9}=1$

The tree diagram itself can be used to calculate the probability of **combined events** too.

To work out the probability of a particular outcome, the probabilities on those branches are multiplied together.

For example, the probability of picking two lemon sweets is

$$P(L,L)=\frac{4}{10}\times\frac{3}{9}=\frac{12}{90}$$

The sum of the probabilities of all four outcomes will equal 1.

Therefore P(M, M) + P(M, L) + P(L, M) + P(L, L)=1.

Worked example

A student goes to school by car (C) 70% of the time and by bicycle (B) the rest of the time.

If she travels by car she arrives on time (Ot) 80% of the time and late (L) the rest of the time.

If she travels by bicycle she is on time 90% of the time and late the rest of the time.

a Draw a fully labelled tree diagram to show both how the student travels to school and whether she arrives on time.

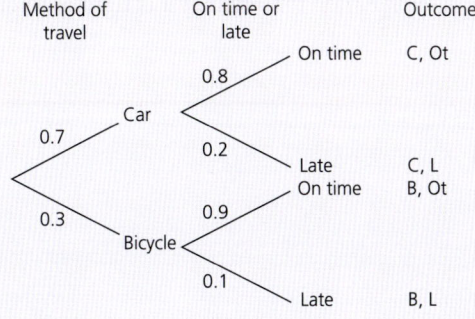

LET'S TALK

How many outcomes show the student arriving to school on time?

KEY INFORMATION

The probabilities should be written as a fraction or decimal rather than as a percentage on the tree diagram and are written on each branch.

b Calculate the probability that the student travels by car and arrives late.

$P(C, L) = 0.7 \times 0.2 = 0.14$

c Calculate the probability that the student arrives at school on time.
There are two branches that show the student arriving on time:

- Car and On time
- Bicycle and On time

The probabilities of each of these are:

$P(C, Ot) = 0.7 \times 0.8 = 0.56$

$P(B, Ot) = 0.3 \times 0.9 = 0.27$

Therefore, the probability of arriving on time is
$0.56 + 0.27 = 0.83$

LET'S TALK

Without repeating the whole calculation, how can you quickly calculate the probability of the student arriving late?

d Are the method of transport and whether the student arrives on time independent events?

No, a student arriving on time or being late is affected by their mode of transport.

Exercise 16.1

1 Below is a list of some pairs of events.
For each pair:

i) decide whether the events are independent or not

ii) if they are dependent, give a reasoned explanation why.

  a Flip a coin and get a head, then flip a second coin and get a head.

  b The probability of it being a cloudy day and the probability of it raining on that day.

  c The probability of someone liking animals and the probability of them having a bird as a pet.

  d The probability of rubbish being collected and the probability of it being a sunny day.

e The probability of it being rainy today and the probability that it will be rainy tomorrow.

f The probability of being vaccinated and the probability of catching diseases.

2 Three friends are playing a board game together. To start they must roll a six with the dice. The first friend rolls a six, and so does the second one. The third friend says 'That's not fair, because now I'm less likely to roll a six.'

Critique the third friend's statement, giving a reasoned mathematical explanation for whether it is right or wrong.

3 A student decides to flip an unbiased coin twice and each time record whether it lands on heads (H) or tails (T).
  a Is the probability of getting a head on the second flip independent of the result of the first flip? Justify your answer.
  b Copy and complete the tree diagram below, adding the probability of each outcome next to each branch.

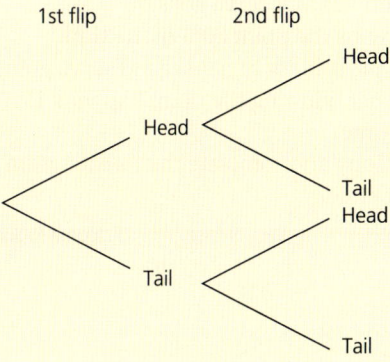

  c **i)** On your diagram highlight the branch(es) representing getting two heads.
  ii) Calculate the probability of getting two heads.
  d Calculate the probability of getting one head and one tail in any order.

4 A circular spinner is divided equally in two, with one half coloured red and the other half green. A second spinner is equally divided into three, with one third coloured green and two thirds blue.
  a Are the spinner results independent of each other? Give a **convincing** reason for your answer.
  b Both spinners are spun. Calculate the probability that:
  i) they both land on green
  ii) they land on different colours.

5 A small child goes to a pet shop that sells tropical fish. In one fish tank there are eight fish. Five of them are red and three are blue.

The child selects two fish at random.
  a Copy and complete the tree diagram below to show the probability of selecting fish of specific colours each time.

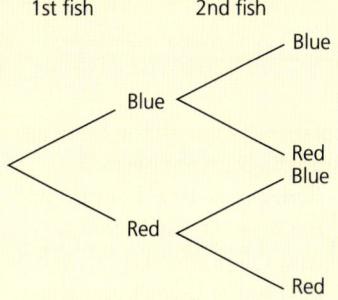

  b Calculate the probability that the child selects two red fish.
  c Calculate the probability that she chooses one fish of each colour.
  d Is the probability of picking a red fish second independent of what colour fish was picked first? Justify your answer.
  e Comment on the validity of the assumption that the child picks the fish at random.

6 A sweet packet contains five coloured sweets. Three are red, one green and one yellow.
A child picks two sweets at random.

a A tree diagram is shown below.

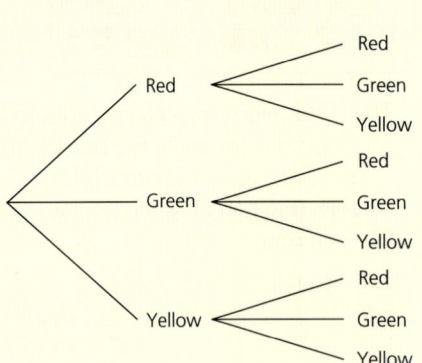

1st sweet   2nd sweet

Explain why not all the branches shown in the diagram are needed.

b i) Redraw the tree diagram with only the necessary branches.

ii) Add the probability of each outcome next to the correct branches.

c Calculate the probability that the child picks two red sweets.

d Calculate the probability that he picks different coloured sweets.

> **LET'S TALK**
>
> Is there an efficient way of calculating the answer to part (d)?

7 A tennis player analyses her serves during a number of matches.
Her serves always go in on either the first or the second serve as she never double faults.
She calculates that 75% of her first serves go in.
When her first serve goes in, she wins the point 80% of the time.
The probability of her second serve going in and her winning the point happens 15% of the time.

> A double fault in tennis occurs when the player hits both the first and the second serves out.

a Copy and complete the tree diagram below.

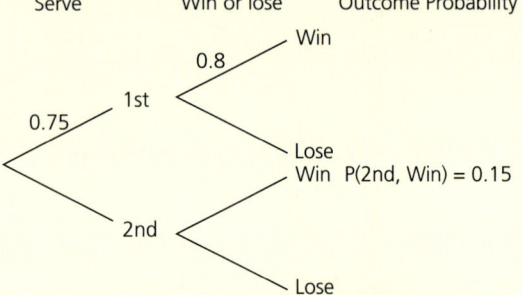

Serve   Win or lose   Outcome Probability

b Calculate the probability that she loses a point.

c Is the probability of winning dependent or independent of the first serve going in? Justify your answer.

KEY INFORMATION

Blaise Pascal (1623–1642) was a genius who studied geometry as a child. He founded probability theory. He is best known for Pascal's triangle which you may already have learned about. If not, then research it to see how it relates to probability.

 Now you have completed Unit 16, you may like to try the Unit 16 online knowledge test if you are using the Boost eBook.

17 Further constructions, polygons and angles

- Construct 60°, 45° and 30° angles and regular polygons.
- Derive and use the formula for the sum of the interior angles of any polygon.
- Know that the sum of the exterior angles of any polygon is 360°.
- Use properties of angles, parallel and intersecting lines, triangles and quadrilaterals to calculate missing angles.

Constructions

You will already know that drawing some triangles accurately involves the use of a pair of compasses. Similarly, you may remember that bisecting an angle or drawing a perpendicular bisector is also done by construction using a pair of compasses.

It is of course possible to construct shapes, bisect angles and perpendicular bisectors using geometry computer programs. You are encouraged to investigate their use whilst doing this section of the book.

Constructing angles

The following are instructions for constructing angles of different sizes using a pair of compasses.

Angle of 60°

- Draw a line, e.g. 5 cm long.
- Keeping the compasses open the same distance, place the compass point on one end of the line and draw an **arc**.

- Place the point on the other end of the line and draw another arc so that it intersects the first one.

- Joining one end of the line to the point of intersection of the two arcs, with a straight line, forms an angle of 60°.

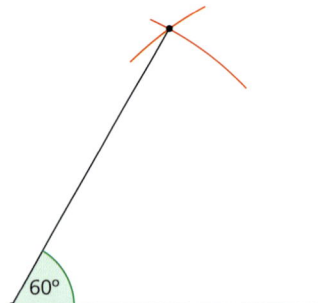

Angle of 30°

This can be constructed by **bisecting** an angle of 60°.

- Open the pair of compasses to, say, 3 cm and place the compass point on the corner of the angle. Draw two arcs so that they intersect the lines as shown.

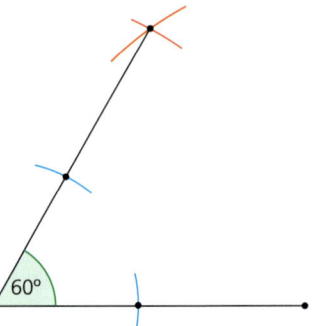

- Place the compass point on each of the two intersections with the lines and draw two further intersecting arcs.

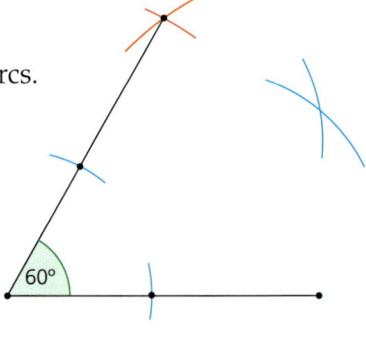

- Drawing a straight line from the corner of the angle through the intersection of these two arcs **bisects** the angle and therefore forms an angle of 30°.

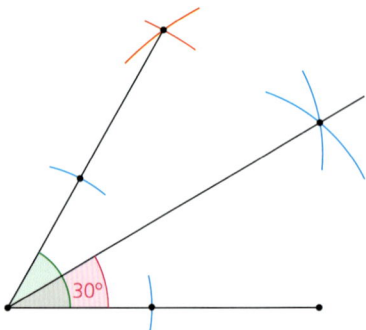

Angle of 45°

This can be constructed by bisecting an angle of 90°.

The angle of 90° is constructed by drawing two lines perpendicular to each other.

- Draw a line, e.g. 5 cm.
- Open your pair of compasses, place the compass point on one end of the line and draw two arcs, above and below the line.

LET'S TALK

What is the minimum distance the pair of compasses can be open to for this construction?

- Place the compass point on the other end of the line and draw two more arcs above and below the line, so that they intersect the first two arcs.

- Draw a straight line through the two points of intersection. This line is a **perpendicular bisector** of the first line.

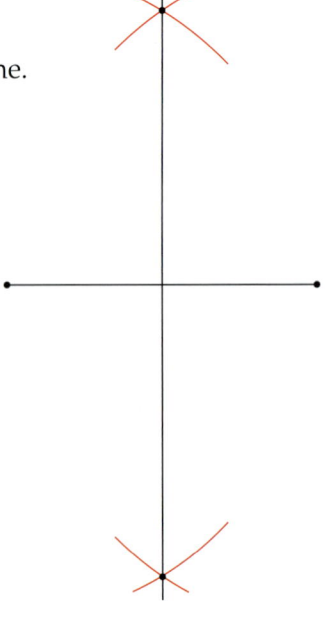

- To construct the 45° angle, this right-angle is bisected using the same method as given earlier for the 30° angle.
  This will look as shown here.

45°

Inscribing polygons

An **inscribed polygon** is one that is drawn inside a circle so that all of its vertices lie on the circumference of the circle. The basic ones can all be constructed using the methods you have used above.

Inscribed hexagon and equilateral triangle

A hexagon or equilateral triangle can be constructed following these steps.

- Open the pair of compasses, e.g. to 4 cm, and draw a circle. With the compasses still open to the same distance, place the point on the circumference of the circle and draw an arc so that it intersects the circumference of the circle.

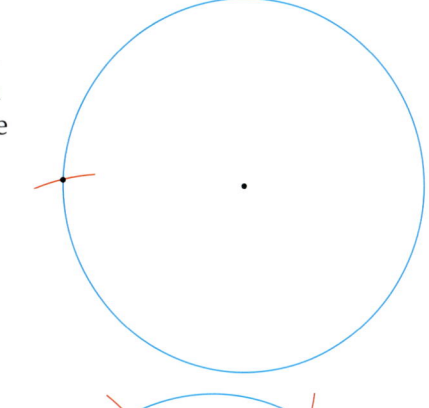

- Place the compass point on the intersection of the arc with the circle and draw another arc so that it too intersects the circle. Repeat this process until the last arc intersects the starting point.

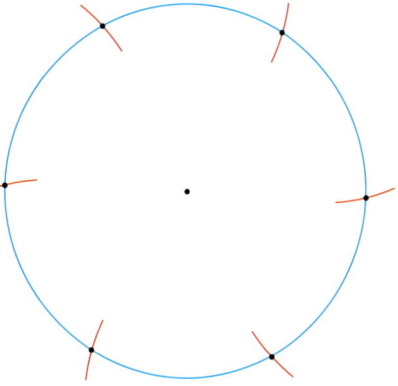

- Joining the adjacent intersection points with straight lines forms a regular hexagon, whilst joining every other intersection point produces an equilateral triangle.

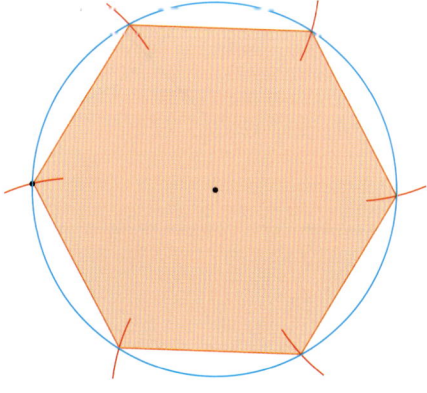 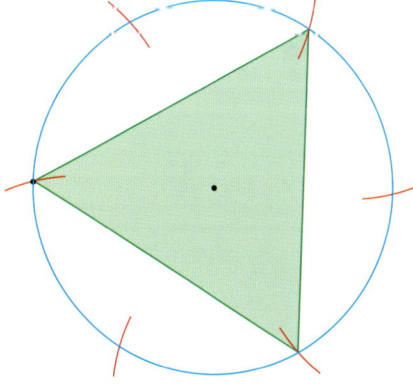

Inscribed square and octagon

Squares and octagons can also be constructed using the methods introduced earlier. Therefore briefer descriptions of the method of construction will be given here in order to encourage you to try and construct them.

- Draw a circle and add a diameter.
- Construct the diameter's perpendicular bisector.
- Join the four points where the diameter and the perpendicular bisector intersect the circle to form a square.

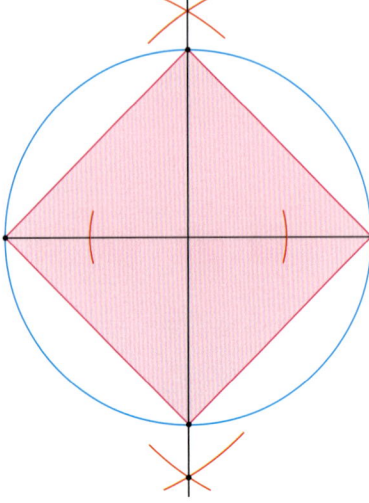

- To construct the octagon, construct the perpendicular bisector of each of the sides of the square. Where these intersect the circle can also be joined to the vertices of the square to form the regular octagon.

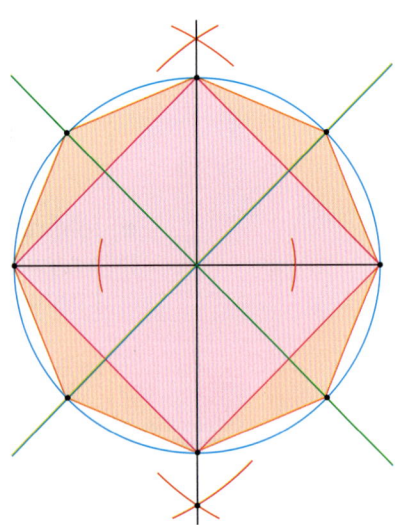

Exercise 17.1

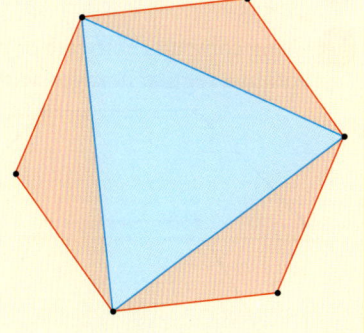

1 Using a pair of compasses and a ruler, construct an equilateral triangle inside a regular hexagon as shown. Leave all construction lines visible.

2 Construct a triangle ABC with AB = 8 cm, angle ABC = 45° and angle BAC = 60°. Leave all the construction lines visible.

3 A student wishes to bisect the angle BAC below.

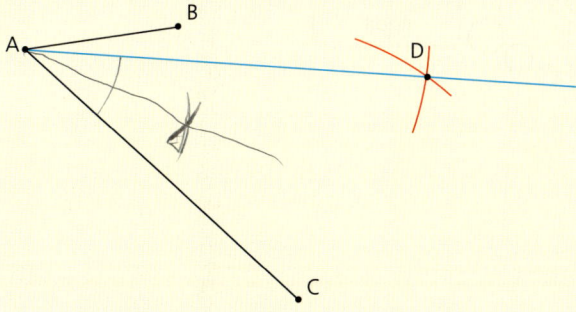

He opens his pair of compasses and, keeping the distance opened the same, places the compass point on B and C in turn and draws an arc each time.
He labels the point of intersection of the two arcs D.
He draws a straight line from A through D, but realises that he has not bisected the angle correctly.

  a Explain clearly the mistake the student has made.

  b By drawing the points A, B and C in approximately the same positions as shown, construct the angle bisector for ∠BAC.

4 It is stated that it is possible to find the centre of an equilateral triangle by constructing the perpendicular bisectors of each of the three sides of the triangle and finding where they intersect.

  a **i)** By constructing an equilateral triangle first, construct perpendicular bisectors for the three sides.

      ii) Do they intersect at the same point?

  b Where is the centre of the circle in relation to any points of intersection of the three perpendicular bisectors?

5 By construction, construct a right-angled isosceles triangle with a base length of 6 cm as shown.

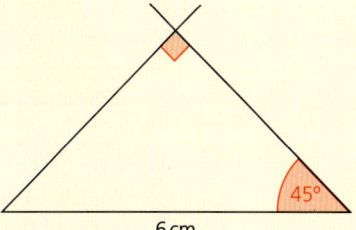

6 cm

6 Construct angles of the following sizes.

    a   135°   b   120°

7 Using computer software or by construction, construct an equilateral triangle within a square, within a regular hexagon within a regular octagon as shown.

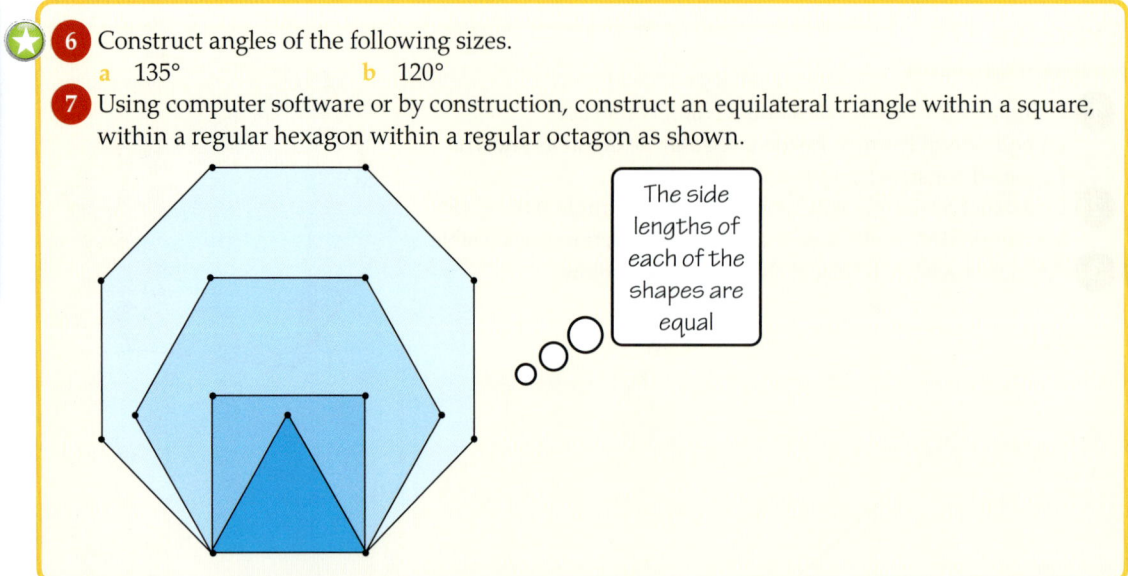

The side lengths of each of the shapes are equal

Polygons

A polygon is a two-dimensional shape with three or more sides, all of which are straight. Examples include triangles, quadrilaterals, pentagons, hexagons etc.

Interior angles of polygons

You will already know that the three **interior angles** of any triangle always add up to 180°. This result can be used to calculate the sum of the interior angles for any polygon.

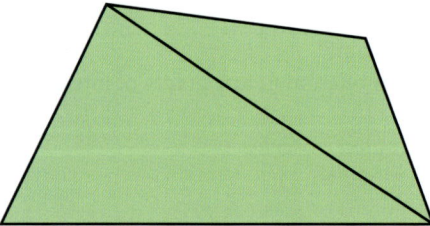

This diagram shows a quadrilateral split into two triangles.

As the sum of the interior angles of each triangle is 180°, then the sum of the four angles of any quadrilateral must be $2 \times 180° = 360°$.

Other polygons can also be split into triangles, as shown.

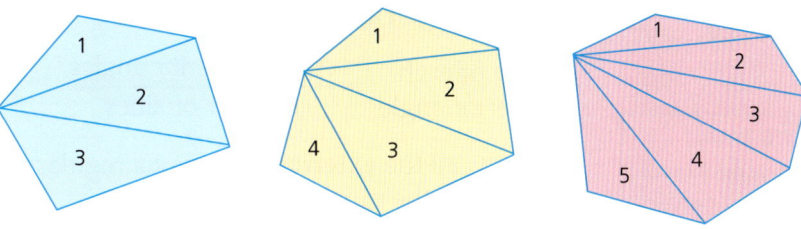

There is a relationship between the number of sides a polygon has and the number of triangles that it can be split up into and this relationship will help deduce how to work out the sum of the interior angles of any polygon.

The diagrams can be summarised as follows:

Number of sides	5	6	7
Number of triangles	3	4	5

The results can be written as a function machine as:

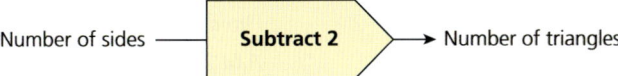

As the sum of angles of a triangle is 180°, then the total of the interior angles of a polygon will equal the number of triangles multiplied by 180.

The function machine can be modified to work out the sum of the interior angles as:

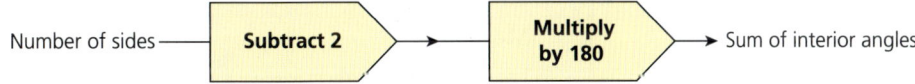

If n represents the number of sides and S the sum of the interior angles, the function machine can be written algebraically as:

$$S = 180(n-2)$$

This formula can be used to find the sum of the internal angles of any polygon. It can also be used to find the size of each internal angle in a regular polygon.

A **regular polygon** is one in which all the sides are the same length and all the angles are equal in size.

Therefore you can calculate the size of each interior angle of a regular polygon as follows:

$$\text{Size of each interior angle} = \frac{\text{sum of interior angles}}{\text{number of sides}}$$

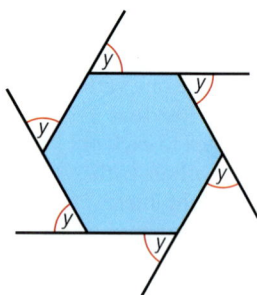

Similarly, all the **exterior angles** of a regular polygon are the same size:

$$\text{Size of each exterior angle} = \frac{360°}{\text{number of sides}}$$

Worked example

a   The sum of the internal angles of a polygon is 1800°.

Calculate the number of sides the polygon has.

The formula $S = 180(n-2)$ can be used and rearranged to find the number of sides n.

$$1800 = 180(n-2)$$
$$10 = n-2 \qquad \textit{Divide both sides by 180.}$$
$$12 = n \qquad \textit{Add 2 to both sides.}$$

Therefore, the polygon has 12 sides.

b   If it is a regular polygon, calculate the size of each internal angle.

12 angles = 1800°

Each angle = 1800 ÷ 12 = 150°

c   If it is a regular polygon, calculate the size of each exterior angle.

There are two ways of calculating this value.

Method 1:

$$\text{Each exterior angle} = \frac{360}{\text{number of sides}}$$
$$= \frac{360}{12}$$
$$= 30°$$

Method 2:

Each exterior angle + each interior angle = 180° as shown.

Exterior angle + 150° = 180°

Exterior angle = 30°

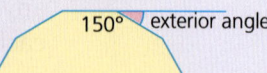

Exercise 17.2

1 By drawing polygons and splitting them into triangles, copy and complete this table.

Number of sides	Name of polygon	Number of triangles	Total sum of interior angles
3	triangle	1	180°
4	quadrilateral	2	$2 \times 180° = 360°$
5	pentagon		
6			
8			
9			
10			
12			

2 Copy and complete this table for regular polygons.

Number of sides	3	4	5	6	8	9	10	12
Sum of the interior angles	180°	360°						
Size of each interior angle	60°							
Size of each exterior angle	120°							

3 The exterior angle of a regular polygon is 2°.
  a Calculate the number of sides the polygon has.
  b Calculate the size of each interior angle.

4 A tile pattern consists of congruent regular octagons and squares arranged as shown. Prove that the tiles tessellate.

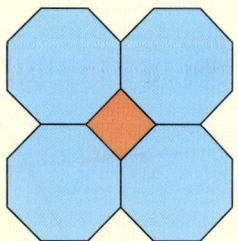

KEY INFORMATION
Shapes that tessellate fit together with no gap.

5 The pattern shows regular dodecagons tessellating with squares and hexagons.

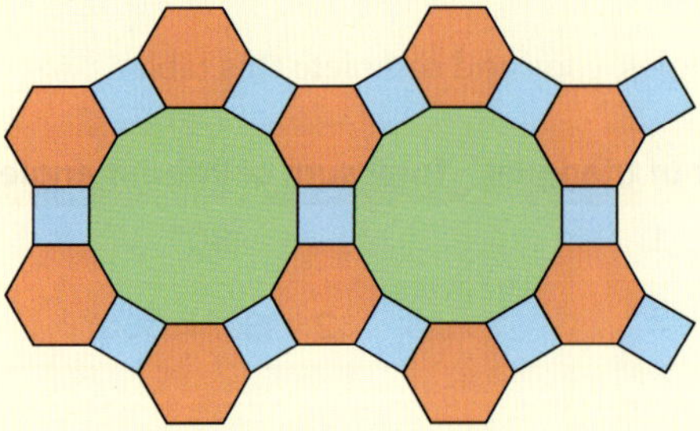

Use mathematics computer programs to design some tessellating patterns of your own.

By looking at the internal angles of these regular polygons, prove that they tessellate.

6 The interior angle of a regular polygon is 19 times the size of the exterior angle. Calculate the size of the interior angle and the number of sides the polygon has.

7 A regular hexagon is drawn inside a regular octagon as shown. Calculate the size of the angle x.

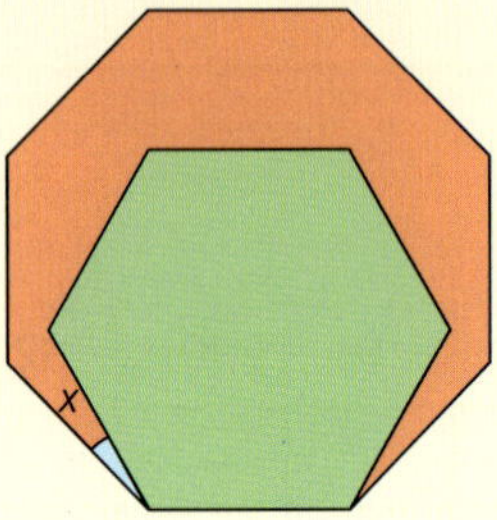

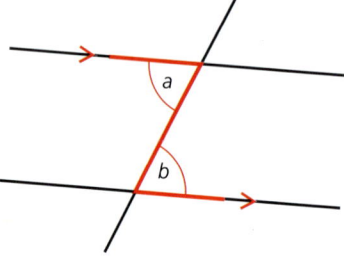

Angle relationships and polygons

In Stage 8 you studied the relationships between angles within parallel and intersecting lines. You know that:

- Alternate angles are equal. They can be found by looking for a 'Z' formation in a diagram.

angle a = angle b

- Corresponding angles are equal. They can be found by looking for an 'F' formation in a diagram.

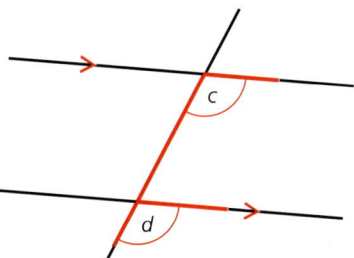

angle c = angle d

> ## Worked example
>
> A pentagon ABCDE is shown below.
>
>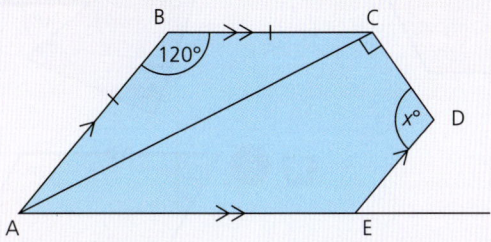
>
> Calculate the size of $x°$, justifying your answer.
>
> $\angle ACB = \angle BAC = 30°$ Isosceles triangle, and angles of a triangle add up to 180°.
>
> $\angle CAE = 30°$ $\angle ACB = \angle CAE$, alternate angles are equal.
>
> Therefore, $\angle BAE = 60°$ and $\angle BCD = 120°$
>
> $\angle DEF = \angle BAE = 60°$ Corresponding angles are equal.
>
> $\angle AED = 120°$ Angles on a straight line add up to 180°.
>
> Interior angles of a pentagon add up to 540°.
>
> Therefore $x = 540 - 120 - 60 - 120 - 120 = 120°$.

Exercise 17.3

Calculate the size of the unknown angles in each of the following polygons.
Give **convincing** reasons for your answers.

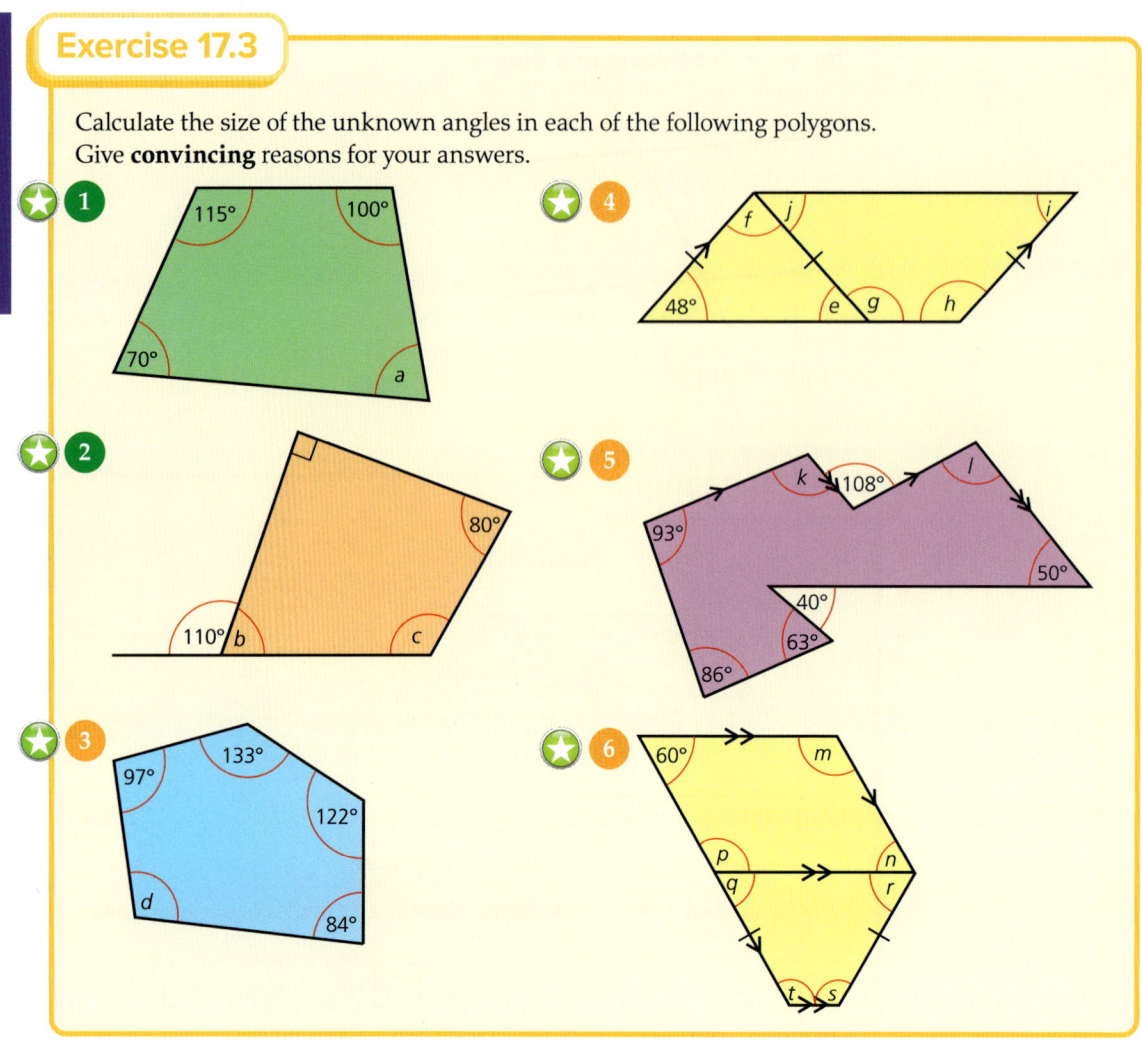

Now you have completed Unit 17, you may like to try the Unit 17 online knowledge test if you are using the Boost eBook.

- Understand that a situation can be represented either in words or as an algebraic expression, and move between the two representations (including squares, cubes and roots).
- Understand that a situation can be represented either in words or as a formula (including squares and cubes), and manipulate using knowledge of inverse operations to change the subject of a formula.

You will already have encountered a lot of situations involving the use of algebra. These included algebraic expressions, formulae and equations.

This unit will look in particular at harder cases of algebraic expressions and formulae, whilst harder equations are covered in more depth in Unit 20 of this book.

Algebraic expressions

Worked examples

1 a Write an expression for the area of the triangle shown.

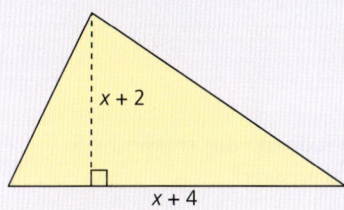

Remember that an algebraic expression is simply a mathematical statement. It has no equals sign.

The formula for the area (A) of a triangle is $A = \frac{1}{2} \times \text{base} \times \text{height}$.

$$A = \frac{1}{2}(x+4)(x+2)$$

$$= \frac{1}{2}\left(x^2 + 6x + 8\right)$$

$$= \frac{1}{2}x^2 + 3x + 4$$

Therefore, the expression for the area is $\frac{1}{2}x^2 + 3x + 4$.

b Calculate the value of the expression when $x = 4$.

Substituting $x = 4$ into $\frac{1}{2}x^2 + 3x + 4$ gives:

$$\frac{1}{2}(4)^2 + 3(4) + 4 = 8 + 12 + 4$$
$$= 24$$

> **LET'S TALK**
>
> Using the diagram, can a value for the area be calculated in a different way?

2 I think of a number n, multiply it by 3, square the answer and then add 6 to it.

a Write an expression for the end result.

$$(3n)^2 + 6$$

> When carrying out calculations, remember to take into account BIDMAS and the order of operations.

b If the original number was 4, evaluate the final answer.

Substituting 4 for n in $(3n)^2 + 6$ gives:

$$(3 \times 4)^2 + 6 = 12^2 + 6$$
$$= 150$$

Exercise 18.1

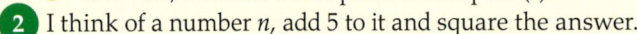

1 A rectangle has dimensions as shown.

> $x + 5$
>
> $x - 1$

    a Write an expression for
        i) its perimeter   **ii)** its area.
    b If $x = 10$, evaluate the expressions in part (a).

2 I think of a number n, add 5 to it and square the answer.

    a Which of the following expressions represent the final result?

$$(n+5)^2 \qquad n+5^2 \qquad n^2+5^2 \qquad n^2+10n+25$$

    b Justify your choice(s) in part (a).
    c If $n = 8$, evaluate the final answer.

3 I think of a number x, double it, square the result and then divide the answer by 5.

    a Which of the following expressions represent the final answer?

$$\frac{2x^2}{5} \qquad \frac{(2x)^2}{5} \qquad \frac{4x^2}{5} \qquad \frac{4}{5}x^2$$

    b Justify your choice(s) in part (a).
    c If $x = 15$, evaluate the final answer.

4 A square has side length a.

    a Write expressions for:
        i) its perimeter
        ii) its area
        iii) the length of its diagonal.
    b If $a = 5$, calculate the length of the diagonal.

5 A cube has a side length of $m + 2$.

    a Write an expression for the volume of the cube.
    b Write an expression for the surface area of the cube.

6   A number p is squared and then multiplied by 4.
Then 5 is subtracted from the answer.
   a   Write an expression for the final result.
   b   i)   Evaluate the expression when $p = 8$.
       ii) Evaluate the expression when $p = -8$.
       iii) Comment on any similarities between your answers to parts (i) and (ii).
            Justify your answer.

7   The cube root of a number p is doubled and divided by 3.
Then 8 is subtracted from this answer.
   a   Write an expression for the final answer.
   b   If $p = 27$, calculate the value of the final answer.

8   A student is showing his work to his teacher.
He is trying to write an expression for the length of
the diagonal of the rectangle shown.
He has written the following working:

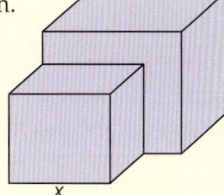

   Using Pythagoras' theorem:
   $H^2 = x^2 + 4x^2$
   $H^2 = 5x^2$
   $H = \sqrt{5x^2}$

   Therefore H is given by the expression $\sqrt{5}x$
The teacher says that the student has made a mistake.
   a   Find the mistake and explain what the student has done wrong.
   b   Write the correct expression for H.

9   A cube of side length x is stuck to another of side length y as shown.
Show that the total surface area of the combined shape is given by
the expression $6y^2 + 4x^2$.

10 A regular octagon of side
length x is shown.
Write expressions for:
   a   its perimeter
   b   its area.

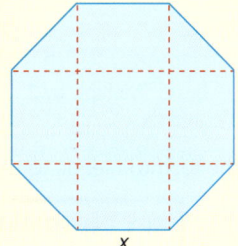

Formulae

A formula is different from an expression in that, rather than simply
being an algebraic statement, it shows the relationship between
variables. As a result it contains an equals (=) sign.

You will already have come across a number of different formulae. These include:

$$A = \pi r^2 \qquad A = \frac{1}{2}bh \qquad A = \frac{1}{2}h(a+b) \qquad C = \pi D \qquad V = l^3$$

In each case the letters represent different variables and the formula shows the way the variables are related.

In the way they have been written above, one letter is on one side of the equals sign on its own. This letter is known as the **subject of the formula**.

However, often formulae need to be rearranged in order to make a different variable the subject of the formula. Being able to do this is an essential mathematical skill.

Worked example

A formula linking the variables a, b and c is given as $a = \dfrac{b^2 - c}{2}$.

a Rearrange the formula to make b the subject.

To answer this it is often useful to look at the variable required and, using BIDMAS, work out which mathematical operations are being used.

In this case a function machine would look as follows:

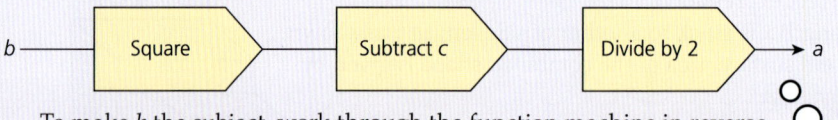

b — Square — Subtract c — Divide by 2 — → a

A function machine is not usually needed. It is used here to demonstrate the order of operations.

To make b the subject, work through the function machine in reverse and carry out the inverse operations, i.e.

b ← Square root — Add c — Multiply by 2 — a

Applying these operations to both sides of the formula leads to the following:

$$a = \frac{b^2 - c}{2}$$
$$2a = b^2 - c \qquad \textit{Multiply by 2.}$$
$$2a + c = b^2 \qquad \textit{Add c.}$$
$$\sqrt{2a + c} = b \qquad \textit{Square root.}$$

> **KEY INFORMATION**
>
> The subject of the formula is on its own on one side with everything else on the other side. The subject can be on its own on the left or on the right.

b Calculate the value of b when $a = 9$ and $c = 7$.

$$b = \sqrt{2 \times 9 + 7} = \sqrt{25}$$
$$= 5$$

Exercise 18.2

1 Rearrange each of the following formulae to make the variable in red the subject.

a $a = 2b + c$

c $m = 5(n + 6)$

e $a = bc - d$

g $r = \frac{1}{2}\sqrt{s}$

b $x = \frac{y}{2} - 4$

d $p = q^2 - 9$

f $m = 2n - p$

h $y = \frac{\sqrt[3]{x}}{2}$

2 Euler's formula gives the relationship between the number of faces (F), vertices (V) and edges (E) of a polyhedron.
It states that $F + V - E = 2$.

a i) Rearrange the formula to make V the subject.

ii) Use the formula to calculate the number of vertices a polyhedron has when $E = 9$ and $F = 5$.

iii) Sketch a possible polyhedron with these properties.

b A polyhedron has 10 faces and 16 vertices.
Give a reasoned argument why it cannot have 20 edges.

> **KEY INFORMATION**
> A polyhedron is a 3D shape in which all of the faces are polygons.

3 A cuboid has dimensions as shown.

a Write a formula in terms of l for the volume (V) of the cuboid.

b i) Rearrange the formula to make l the subject.

ii) Calculate the length of each of the sides when $V = \frac{1331}{25}$ cm^3.

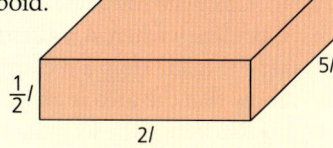

4 a Explain why the volume (V) of a cylinder is given by the formula $V = \pi r^2 l$, where r is the radius of the constant cross-section and l is the length.

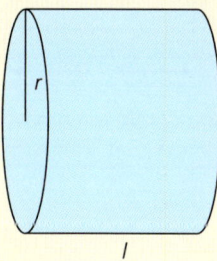

b Calculate the volume when $r = 6$ cm and $l = 12$ cm.

c i) Rearrange the formula to make l the subject.

ii) Calculate the length when $V = 3000$ cm^3 and $r = 4$ cm.

d Calculate the value of r when $V = 5000$ cm^3 and $l = 50$ cm.

5 Rearrange each of the following formulae to make the variable in red the subject.

a $a = \frac{2b - c}{5}$

c $m = \sqrt{\frac{n - 6}{3}}$

e $c = \frac{d^{\frac{1}{3}}}{5}$

b $x = h^2 - 4$

d $p = 8q^3$

f $x = \frac{4 - t^2}{y}$

6 **a** In the rectangle shown explain why $H^2 = x^2 + y^2$.

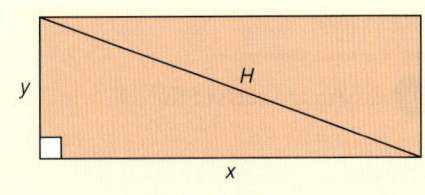

  b **i)** Rearrange the formula to make x the subject.

  ii) Calculate the value of x if $y = 36$ and $H = 60$.

7 The formula for the area of a trapezium

  is given as $A = \dfrac{h}{2}(a+b)$.

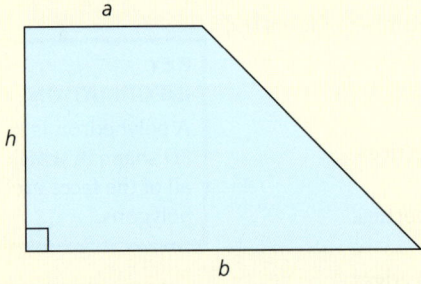

  a **i)** Rearrange the formula to make a the subject.

  ii) Calculate the value of a when $A = 120\,\text{cm}^2$, $h = 4\,\text{cm}$ and $b = 50\,\text{cm}$.

  b Calculate h when $A = 40\,\text{cm}^2$, $a = 8\,\text{cm}$ and $b = 12\,\text{cm}$.

8 A shape is made from a rectangle and two semicircles as shown.

  a Show that the area is given by the formula $A = \pi r^2 + 2rl$.

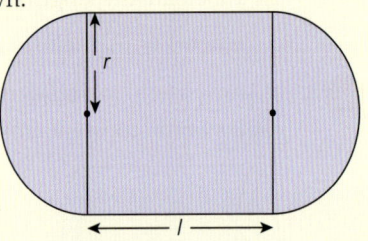

  b **i)** Rearrange the formula to make l the subject.

  ii) Calculate the value of l when $A = 200\,\text{cm}^2$ and $r = 5\,\text{cm}$.

  c **i)** If $r = l$ simplify the formula for the area, giving your answer in terms of r and in factorised form.

  ii) Calculate the radius if $A = 500\,\text{cm}^2$ when $r = l$.

9 Three cubes are attached as shown.

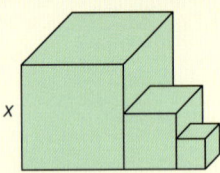

The length of each edge of the largest cube is x cm.

The lengths of the edges of the following cubes are half that of the previous one.

  a Write a formula for the total volume, V, of the three cubes in terms of x. Give your answer as a fraction in its simplest form.

  b **i)** Rearrange the formula to make x the subject.

  ii) Calculate the total length of the bottom edges of the three cubes together when $V = 400\,\text{cm}^3$. Give your answer correct to 1 d.p.

10 A regular hexagon has a side length of x and a height h as shown.

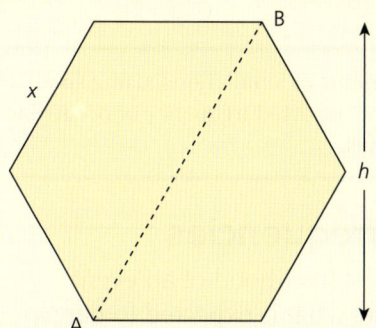

KEY INFORMATION

A **geometric proof** uses the shape's properties to prove the result.

a   Write the length AB in terms of x.
b   Show that $h = \sqrt{3}x$.
c   i)   Rearrange the formula to make x the subject.
     ii) Calculate the value of x when $h = 6\,\text{cm}$.
d   Prove geometrically or algebraically that the area of the

hexagon is given by the formula $A = \dfrac{3\sqrt{3}}{2}x^2$.

LET'S TALK

Can you prove the formula for the area using both types of proof?

 Now you have completed Unit 18, you may like to try the Unit 18 online knowledge test if you are using the Boost eBook.

- Design and conduct chance experiments or simulations, using small and large numbers of trials. Calculate the expected frequency of occurrences and compare with observed outcomes.

Expected and relative frequencies

You will know that probability is the likelihood of an event happening. As it deals with chance, what is expected to happen (the **theoretical probability**) does not necessarily happen. This unit will look in more detail at what happens to observed results (the **relative frequency**) compared with the theoretical probability and therefore the **expected frequency**, as an experiment is repeated many times.

For an unbiased coin, as the two outcomes, heads and tails, are equally likely it can be stated that the probability of getting a tail is equal to half, so:

> Note the different use of the word 'frequency'.

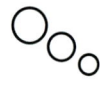

$$P(\text{Tail}) = \frac{1}{2} = 0.5$$

This is the **theoretical probability**.

Expected frequency refers to the actual number of times an outcome is expected to occur, whilst the relative frequency is a probability. If the coin is flipped 50 times, the expected frequency of tails would be 25.

A student flips an unbiased coin five times and records the number of tails. She gets 'tails' four times. This figure is the observed result of the experiment. From this it can be stated that

LET'S TALK

Does getting 'tails' four times out of five imply that the coin is biased?

$$P(\text{Tail}) = \frac{4}{5} = 0.8$$

This is the **relative frequency**.

LET'S TALK

Does getting 'tails' seven times out of ten imply that the coin is biased?

The student flips the coin another five times and gets tails three times. As the results have changed, this means that the relative frequency has also changed. So far the student has recorded 'tails' seven times out of ten flips of the coin.

The relative frequency is now $P(\text{Tail}) = \frac{7}{10} = 0.7$

A formula for **expected frequency** can be given as:

Expected frequency = theoretical probability × number of trials

A formula for the relative frequency can be given as:

$$\text{Relative frequency} = \frac{\text{observed frequency}}{\text{number of trials}}$$

The student carries out the experiment a total of ten times and records the number of tails she gets out of each set of five flips. The results are shown below.

Experiment	1	2	3	4	5	6	7	8	9	10
Number of tails	4	3	3	1	2	1	3	2	3	2

To see how the relative frequency changes as the number of times the experiment is carried out increases, further rows can be added to the table.

Experiment	1	2	3	4	5	6	7	8	9	10
Number of tails	4	3	3	1	2	1	3	2	3	2
Cumulative number of tails	4	7	10	11	13	14	17	19	22	24
Cumulative number of coin flips	5	10	15	20	25	30	35	40	45	50
Relative frequency	0.80	0.70	0.67	0.55	0.52	0.47	0.49	0.48	0.49	0.48

KEY INFORMATION

Cumulative means the running total.

LET'S TALK

Which value of relative frequency is a more reliable indicator of theoretical probability?

From the table above it can be seen that the relative frequency of getting a tail changed as the experiment was repeated.

After the first experiment, $P(\text{Tail}) = \frac{4}{5} = 0.8$, whilst by the last experiment, after 50 flips of the coin, $P(\text{Tail}) = \frac{24}{50} = 0.48$

A graph of the results looks like this:

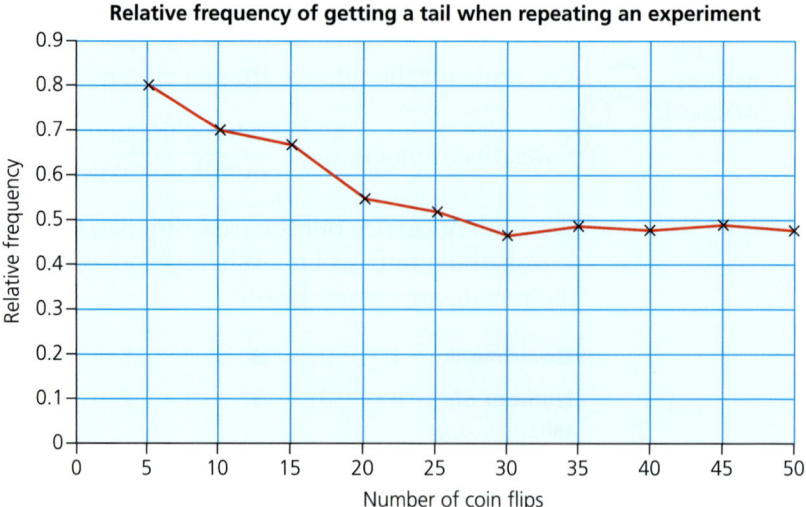

Relative frequency of getting a tail when repeating an experiment

LET'S TALK

Why are you more likely to see large changes in the relative frequency at the start of the graph rather than at the end?

It can be seen from the graph that initially the relative frequency changes considerably, as shown by the large drops in the graph. However, by the end, the graph is fairly constant, only varying slightly.

Worked example

A child suspects that a coin he has is biased against heads.

He flips the coin five times and records the number of heads.

He repeats this experiment a total of ten times (i.e. a total of 50 flips of the coin).

He plots a graph of the relative frequency over the course of the experiments, as shown.

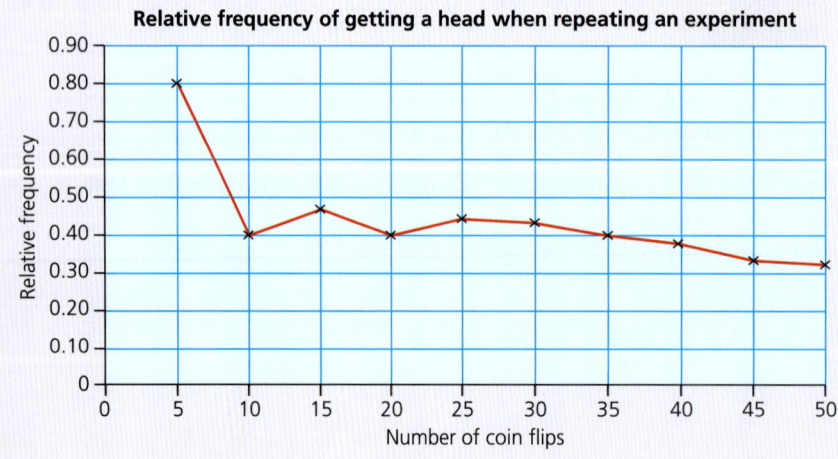

Relative frequency of getting a head when repeating an experiment

a The parent of the child says that the coin appears to be biased towards heads as the relative frequency after five flips is 0.8. Comment on the accuracy of this statement.

Although the relative frequency is 0.8 after five flips, a more accurate assessment of bias can be made after more flips are taken into account. After 50 flips of the coin, the results are showing a relative frequency of getting heads as 0.32.

Therefore, as a normal coin would have a theoretical probability of landing on heads of 0.5, the coin does appear to be slightly biased against heads.

b After 500 flips of the coin, the child works out that the relative frequency of heads is in fact 0.35.

If P(*H*) = 0.35 for the coin, what is the expected frequency of heads after 800 flips?

Expected frequency = theoretical probability × number of trials

Expected frequency = 0.35 × 800 = 280

That is, the child would expect to get approximately 280 heads.

Exercise 19.1

1 A packet contains 15 sweets, 10 of which are a red colour and 5 yellow.
A child decides to conduct a probability experiment. He picks a sweet out at random, notes its colour and then puts it back in the packet. He does this ten times.
To make the results more reliable, he decides to repeat the whole process a total of 15 times, i.e. he has picked out a total of 150 sweets, one at a time, noted the colour and then put it back in the packet.
A graph of his results is shown below.

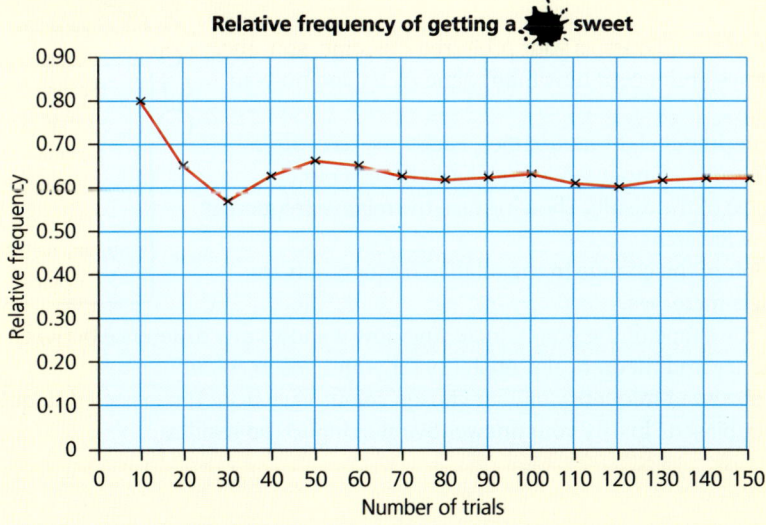

a The title of the graph is partially covered.
Which colour sweet is it likely to be referring to? Justify your answer.

b If the relative frequency of picking that colour of sweet at the end was 0.62, how many times did he pick out that colour of sweet?

2 Two ordinary dice are rolled and their scores are added together.

a Copy and complete the two-way table for the total of the two dice.

		Dice 1					
		1	**2**	**3**	**4**	**5**	**6**
Dice 2	**1**	2					
	2				6		
	3						
	4						
	5				9		
	6						12

b What is the most likely total when two dice are rolled and their scores added?

c **i)** Name a possible outcome that has a theoretical probability of 0.25.

ii) Using two dice, conduct an experiment to calculate the relative frequency of the outcome you chose in part (i). Record your results in a table and show your calculations clearly.

iii) How does the relative frequency compare with the theoretical probability?

3 a Using the construction methods covered in Unit 17, construct a spinner, out of thick card, in the shape of an equilateral triangle. By bisecting each of the angles, find its centre. Colour the spinner using three different colours.

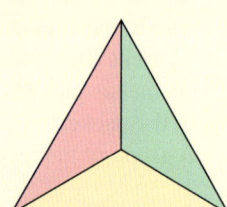

b What is the theoretical probability of each colour?

c If the spinner is spun 90 times, what is the expected frequency of each colour?

d Choosing one of the colours as your preferred outcome, spin the spinner ten times and record how many times it landed on your chosen colour.

By inserting a small pencil through the centre, the triangle can be made into a spinner.

e Repeat the experiment eight more times, each time recording how many times the spinner landed on your chosen colour.

f Construct a table of the results, showing also the relative frequency after each set of ten spins.

g Plot a line graph of the change in the relative frequency as the number of trials increases.

h Comment on the shape of the graph, including how it shows any difference between relative frequency and theoretical probability for your chosen colour.

i How did the observed frequency of your chosen colour vary from the expected frequency?

j Is your spinner biased? Justify your answer by referring to the results.

4 A spinner is split into four congruent sections and numbered 1–4.

   a A student spins it and finds that after ten spins the relative frequency of the spinner landing on the 4 is 0.2. After 100 spins the relative frequency of landing on the 4 is 0.23.
How many times did the spinner land on the 4 between the 10th and 100th spins?

   b Another student uses the spinner.
After 50 spins the relative frequency of landing on the 2 is 0.1.
After 60 spins he calculates the relative frequency of landing on the 2 as $0.2\dot{6}$.
Give a reasoned explanation why this cannot be correct.

5 Place 20 counters of two different colours in a small, non-transparent, bag. The number of each colour can be different.

For experiment 1, remove a counter at random, make a note of its colour and then place it back in the bag. Do this five times.

Conduct the experiment ten times (i.e. a counter has been chosen 50 times in total), each time making a note of the colour of the counter chosen.

   a Construct a table showing the relative frequency of each colour over the ten experiments.

   b **i)** On the same axes plot a line graph for the relative frequency of each colour over the ten experiments.

      ii) Describe the shape of the graphs in relation to each other.

      iii) Justify your findings in part (ii).

Now you have completed Unit 19, you may like to try the Unit 19 online knowledge test if you are using the Boost eBook.

20 Further algebraic equations and inequalities

- Understand that a situation can be represented either in words or as an equation. Move between the two representations and solve the equation (including those with an unknown in the denominator).
- Understand that a situation can be represented either in words or as an inequality. Move between the two representations and solve linear inequalities.

In Unit 18 you studied further cases of algebraic expressions and formulae. This unit will focus on similar cases but involving equations and inequalities.

LET'S TALK

This equation only has one solution, $x = 4$. Can an equation have more than one solution? If so, when?

Recap

An equation is a mathematical statement that shows that two expressions are equal to each other, e.g. $3x - 7 = x + 1$.

The above equation states that the expression $3x - 7$ is equal to the expression $x + 1$. Solving an equation is finding the value(s) of the unknown variable that makes the left-hand side of the equation equal to the right-hand side. In this case the solution is $x = 4$.

An inequality has similarities to an equation, in that it shows a relationship between expressions. However, rather than there being a fixed number of solutions, there is often a range of possible values for the solution. For example, $3x - 7 < x + 1$ states that $3x - 7$ is less than $x + 1$. The solution to this is $x < 4$.

Equations

Solving an equation involves rearranging it to make the unknown variable the subject of the equation. To do this, one fundamental rule should be followed:

Whatever is done to one side of the equation must also be done to the other side.

Worked examples

1 Solve the equation $\dfrac{3x}{2} + 1 = 2x - 2$.

$$3x + 2 = 4x - 4 \qquad \textit{Multiply both sides by 2.}$$
$$2 = x - 4 \qquad \textit{Subtract 3x from both sides.}$$
$$6 = x \qquad \textit{Add 4 to both sides.}$$

2 Solve the equation $\dfrac{9}{x+1} = 3$.

$$9 = 3(x + 1) \qquad \textit{Multiply both sides by } (x + 1).$$
$$3 = x + 1 \qquad \textit{Divide both sides by 3.}$$
$$2 = x \qquad \textit{Subtract 1 from both sides.}$$

> **LET'S TALK**
>
> Could we have subtracted 1 from both sides first? Or subtracted $\dfrac{3x}{2}$ from both sides first?
>
> What different ways are there of solving this equation? Which are easier to do?

3 At the beginning of a party there are x children. By the end of the party there are six more than there were at the start.

At the end, the party host distributes 154 sweets equally among the children. They each get 11 sweets.

How many children were at the start of the party?

An equation can be formed from this information.

The number of children at the end of the party can be expressed as $x + 6$.

154 sweets shared equally between $x + 6$ children can be expressed as $\dfrac{154}{x+6}$

As this is equal to 11 sweets, the equation to be solved is $\dfrac{154}{x+6} = 11$:

$$\dfrac{154}{x+6} = 11$$
$$154 = 11(x + 6) \qquad \textit{Multiply both sides by } (x + 6).$$
$$14 = x + 6 \qquad \textit{Divide both sides by 11.}$$
$$8 = x \qquad \textit{Subtract 6 from both sides.}$$

Therefore, there were 8 children at the start of the party.

Exercise 20.1

1 Solve the following equations.

a $\dfrac{24}{x} = 8$

b $\dfrac{32}{m} + 2 = 10$

c $\dfrac{15 - m}{6} = 4$

d $\dfrac{8}{p} + 6 = \dfrac{2}{p}$

e $\dfrac{14}{2y} = 28$

f $2 - \dfrac{12}{6t} = 0$

2 A student is solving the following equation: $\dfrac{11x}{2x+1} = 5$

He writes the following steps and explanations to solve it:

$$\frac{11x}{2x+1} = 5$$

Step 1: $\dfrac{11x}{2x+1} - 5 = 0$ Subtract 5 from both sides

Step 2: $11x - 5 = 0$ Multiply both sides by $(2x+1)$

Step 3: $11x = 5$ Add 5 to both sides

Step 4: $x = \dfrac{5}{11}$ Divide both sides by 11

a   Explain what mistake the student has made.
A second student also attempts to solve the same equation and writes the following:

$$\frac{11x}{2x+1} = 5$$

Step 1: $11x = 5 \times 2x + 1$ Multiply both sides by $2x+1$

Step 2: $11x = 10x + 1$ Simplify the equation

Step 3: $x = 1$ Subtract $10x$ from both sides

b   **Critique** the second student's method, giving a reasoned justification of its accuracy or error.

c   Solve the equation, showing your method clearly.

3 Solve the following equations.

a $\dfrac{5}{x+1} = 1$

c $\dfrac{1}{2x-5} = 3$

e $2 = \dfrac{12}{1-x}$

b $\dfrac{4x}{x+3} = 3$

d $\dfrac{5x+1}{4x-2} = 3$

f $3 = \dfrac{3(x+4)}{2-3x}$

4 In each of the following:
i)   form an equation from the information given
ii) solve the equation.

a   The perimeter of a regular octagon is 72 cm.
Each side has a length of $(4x-1)$ cm.
Calculate the value of x.

b   A square has side lengths as shown.
Calculate the value of x.

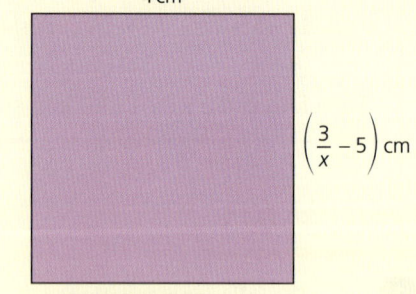

4 cm

$\left(\dfrac{3}{x} - 5\right)$ cm

5 A right-angled trapezium is shown.
If the area of the trapezium is 52 cm² calculate:

a   its height
b   the length of its top edge
c   the length of the sloping edge correct to 3 s.f.

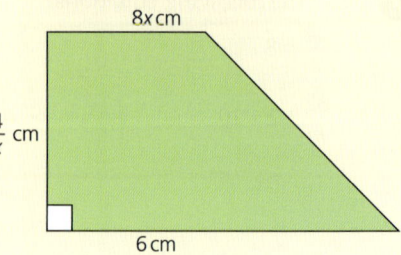

8x cm

$\dfrac{4}{x}$ cm

6 cm

Inequalities

Solving inequalities can be carried out in a similar way to solving an equation in that what is done to one side of the inequality must also be done to the other.

For example, solve $3x + 5 \geqslant 14$

$$3x \geqslant 9 \quad \textit{Subtract 5 from both sides.}$$
$$x \geqslant 3 \quad \textit{Divide both sides by 3.}$$

Therefore, as long as $x \geqslant 3$ then $3x + 5 \geqslant 14$.

This solution can also be represented on a number line as:

KEY INFORMATION

● ——→ A closed circle implies the number is included in the solution and is used with ⩽ or ⩾ inequalities.

○ ——→ A hollow circle implies the number is not included in the solution and is used with < or > inequalities.

When solving an inequality, care must be taken when either multiplying or dividing by a negative number.

If we look at the inequality $4 > 2$ this is a true statement, i.e. 4 is greater than 2.

If the same number is added or subtracted from both sides or if both sides are multiplied or divided by a positive number, then the inequality is still true.

However, if both sides are either multiplied or divided by a negative number then the inequality is not true, e.g. multiplying both sides by −2 produces $-8 > -4$, which is incorrect.

Therefore, when multiplying or dividing by a negative number, the inequality symbol must be reversed to make it true, i.e. $-8 < -4$.

Worked examples

1 a Solve the following inequality: $6 - \dfrac{3x}{2} \geqslant 12$

$$6 - \frac{3x}{2} \geqslant 12$$

$$-\frac{3x}{2} \geqslant 6 \quad \textit{Subtract 6 from both sides.}$$
$$-3x \geqslant 12 \quad \textit{Multiply both sides by 2.}$$
$$x \leqslant -4 \quad \textit{Divide both sides by −3 and switch the inequality sign around.}$$

b Represent the solution on a number line.

$$-6 \quad -5 \quad -4 \quad -3$$

c Check that the solution is correct.

This can be done by substituting any number which satisfies the solution into the original inequality to see if it is correct, e.g. as the solution is $x \leqslant -4$, test $x = -6$.

> Substituting a number greater than -4 would lead to an incorrect statement at the end.

Substituting $x = -6$ into the original inequality gives:

$$6 - \frac{3(-6)}{2} \geqslant 12$$

$$6 - \frac{-18}{2} \geqslant 12$$

$$6 - -9 \geqslant 12$$

$$6 + 9 \geqslant 12$$

$$15 \geqslant 12$$

As 15 is greater than 12, the solution is likely to be correct.

2 **a** The area of the rectangle shown is greater than $42\,\text{cm}^2$ but less than or equal to $48\,\text{cm}^2$. Calculate the possible values for x.

6 cm

$\left(\frac{x}{2} + 4\right)\text{cm}$

The area is $6\left(\frac{x}{2} + 4\right) = (3x + 24)\,\text{cm}^2$

As a result of the information given, the area can be written as an inequality as:

$$42 < 3x + 24 \leqslant 48$$

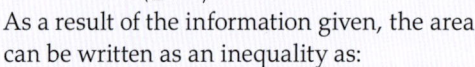

$18 < 3x \leqslant 24$ *Subtract 24 throughout.*

$6 < x \leqslant 8$ *Divide by 3 throughout.*

b Represent the solution on a number line.

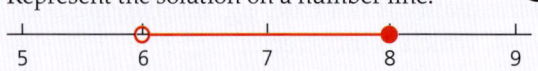

$$5 \quad 6 \quad 7 \quad 8 \quad 9$$

> $42 < 3x + 24 \leqslant 48$ can be split into the two separate inequalities $42 < 3x + 24$ and $3x + 24 \leqslant 48$ and solved separately. However, it is quicker to treat it as a single inequality and apply each calculation to all three parts of the inequality.

c Check that the solution is correct.

The solution has been given as $6 < x \leqslant 8$, therefore test a number within this range, e.g. $x = 7$.

Substituting $x = 7$ into $42 < 3x + 24 \leqslant 48$ gives

$$42 < 3(7) + 24 \leqslant 48$$

$$42 < 45 \leqslant 48$$

As 45 lies between 42 and 48, the solution is likely to be correct.

Exercise 20.2

1 **i)** Solve the following inequalities.
ii) Express the answer on a number line.

a $\dfrac{4x}{5} \le 12$ c $3(x-4) < 2x+1$ e $3 \le 2x-5 < 7$

b $\dfrac{2x-5}{3} > -3$ d $4 < x+7 \le 9$ f $7 \le 3x+5 < 11$

2 In each of the following:
i) form an inequality from the information given
ii) solve the inequality to find the answer.

a Merlin thinks of a whole number. He doubles it and subtracts 8. His answer is greater than 16 but less than 20. What number(s) could he possibly have chosen?

b A taxi fare is calculated as $6 plus $1.50 for each kilometre travelled. A customer estimates that the cost for her journey home will be between $48 and $51. Calculate the range of distances her home must be from where she is.

c Ruth thinks of a number, multiplies it by 3 and subtracts the result from 10. She gets a final answer between −8 and 1. What range of numbers does Ruth's initial number lie in?

3 Two function machines P and Q are shown.

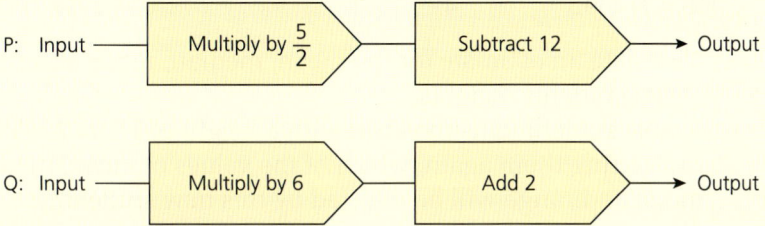

Both function machines have the same input.
Calculate the range of possible input values if the output of P is greater than or equal to the output of Q.

4 A farm is building a small rectangular enclosure for a chicken with dimensions as shown.
The area of the enclosure must be greater than $15\,\text{m}^2$ but less than or equal to $63\,\text{m}^2$.
The perimeter of the enclosure must be at least $32\,\text{m}$ but less than $36\,\text{m}$.
Calculate the possible range of values for p.

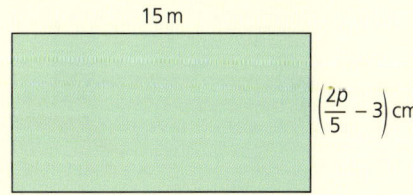

Now you have completed Unit 20, you may like to try the Unit 20 online knowledge test if you are using the Boost eBook.

Section 2 - Review

1 A lorry driver knows that the height of his vehicle is 4.9 m correct to two significant figures. A road he is travelling on has a bridge with a height above the road of 4.92 m.

   a Is it safe for the lorry driver to drive under the bridge? Justify your answer.

   b He then realises that the height of his vehicle is actually 4.90 m correct to three significant figures. Is it safe for the lorry driver to drive under the bridge now? Justify your answer.

2 The heights (H cm) of students in a class are given in the grouped frequency table on the right.

Height (cm)	Frequency
$150 \leqslant H < 160$	6
$160 \leqslant H < 170$	9
$170 \leqslant H < 180$	6
$180 \leqslant H < 190$	4

   a Calculate an estimate for the mean, median, mode and range of the students' heights.

   b A new student joins the class with a height of 172 cm. Without calculating them again, which of the values of mean, median, mode and range will be affected by this new student? Give a reasoned explanation for your answer.

3 A triangle ABC is enlarged by a scale factor of 3 from a centre of enlargement O, as shown.

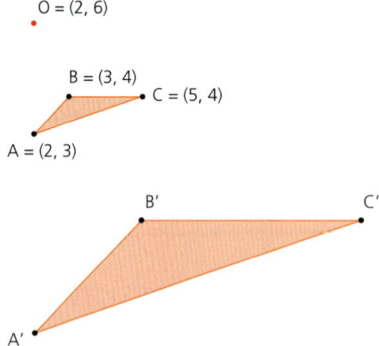

Calculate the coordinates of the vertices of the enlarged shape A′B′C′.

4 Without a calculator, work out the answer to the following calculations. Show your working clearly.

   a $1\frac{3}{4}-\left(\frac{2}{3}-\frac{1}{5}\right)$

   b $2\frac{5}{8}+\left(\frac{5}{3}-1\frac{1}{6}\right)$

5 A shape has dimensions as shown.

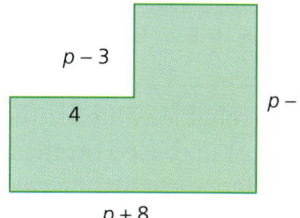

Write an expression for the area of the shape.
Give your answer in its simplest form.

6 In a particular seaside town, the probability of sunshine or rain is calculated.
   On average, the probability of it being a sunny day is 0.7, whilst the probability of rain is 0.3. However, if it is sunny one day, the probability of it being sunny the next is 0.85. If it rains on one day, then the probability of it raining the next day is 0.4.
   A family travels to the seaside town for a weekend break.

   a Draw a fully labelled tree diagram to show all the possible outcomes over the two days.

   b i) What is the probability that the family has two sunny days?
      ii) What is the probability that the family has at least one rainy day?

7 a Using a pair of compasses, construct an equilateral triangle.

   b Using your equilateral triangle, construct one 30° angle.

8 I think of a number p. I subtract it from 6 and then square the answer.

   a Which of the following expressions represents the final result?

      $p^2 - 12p + 36$ $(p-6)^2$ $(6-p)^2$ $6^2 - p^2$

   b If $p = -2$, evaluate the answer.

9 An experiment involves dropping a drawing pin ten times to see how many times it lands 'point up'. The experiment is then carried out a further nine times (i.e. 100 drops in total) and the results presented in the table below.

Experiment	1	2	3	4	5	6	7	8	9	10
Number of times pin landed 'point up'	8	6	6	7	7	5	8	7	6	6

  a Calculate the most reliable value for the relative frequency of the pin landing point up based on these results.
  b Assume that the relative frequency calculated in part (a) is an accurate value for the theoretical probability of the pin landing point up.
     What is the expected frequency of the pin landing point **down** if it is dropped 450 times?
10 A rectangle has a length of $(m - 2)$ cm and width of 8 cm. If its area is at least 92 cm^2 but less than 108 cm^2, calculate the possible range of values for m.
     Give your answer as an inequality.

SECTION 3

History of mathematics – The French

"Mathematics is the music of reason."

John Sylvester

In the middle of the 17th Century there were three great French mathematicians, René Descartes, Blaise Pascal and Pierre de Fermat.

René Descartes (1596–1650) wrote the book *The Meditations,* which asks 'How and what do I know?'

His work in mathematics linked algebra and geometry. He thought that all nature could be explained in terms of mathematics.

Blaise Pascal (1623–1662) studied geometry as a child. At the age of sixteen he stated and proved Pascal's theorem, which connects any six points on any cone. The theorem is sometimes called the 'Cat's Cradle'. He founded probability theory and worked with calculus. He is best known for Pascal's triangle.

Pierre de Fermat (1607–1665) was a brilliant mathematician. He invented number theory (look it up!) and worked on calculus. With Pascal he discovered probability theory.

Fermat is known for 'Fermat's last theorem'. This theorem is derived from Pythagoras' theorem, which proves that in a right-angled triangle of sides x, y and z, then $x^2 + y^2 = z^2$.

Fermat said that if the index was greater than 2 and x, y and z are all whole numbers then the equation was never true. This theorem was only proved in 1995 by the English mathematician Andrew Wiles.

▲ René Descartes

▲ Blaise Pascal

21 Linear and quadratic sequences

- Generate linear and quadratic sequences from numerical patterns and from a given term-to-term rule (any indices).
- Understand and describe nth term rules algebraically (in the form $an \pm b$, where a and b are positive or negative integers or fractions, and in the form $\frac{n}{a}$, n^2, n^3 or $n^2 \pm a$, where a is a whole number).

Term-to-term rules

Recap

A sequence of numbers follow a pattern or a rule. The sequence can be described in one of two ways: a term-to-term rule and a rule for the nth term.

> **KEY INFORMATION**
>
> A term refers to each number in the sequence. A term-to-term rule explains how to get from one term to the next.
> The rule for the nth term explains how to get from its position in the sequence to the term itself.

Position	1	2	3	4	5
Term	2	5	8	11	14

Here the sequence is 2, 5, 8, 11, 14. It can be described using the term-to-term rule, as:

The first term is 2 and the term-to-term rule is add 3.

> **LET'S TALK**
>
> Why is the first term needed for the term-to-term rule?

It can also be described using the formula for the nth term, as $3n - 1$.

> **LET'S TALK**
>
> Which rule would be more useful for calculating the 100th term? Why?

Up until now the sequences you have encountered have involved linear rules. This unit will therefore focus on sequences with more complicated rules. These may include quadratic and even cubic rules.

Worked examples

1   The first term (t_1) of a sequence is 2 and the term-to-term rule is square the term and then subtract 4.

Write down the first five terms of the sequence.

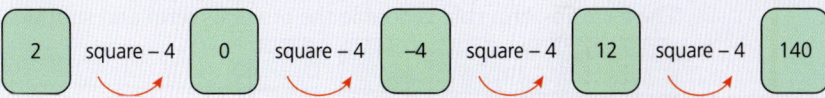

Therefore, the first five terms of the sequence are 2, 0, −4, 12, 140.

Another way to represent this information is:

$t_1 = 2$

$t_2 = 2^2 - 4 = 0$

$t_3 = 0^2 - 4 = -4$

$t_4 = (-4)^2 - 4 = 12$

$t_5 = 12^2 - 4 = 140$

2   Bacteria are left to grow in a petri dish over a period of time (days).

After 1 day there are estimated to be only 2 bacteria. The term-to-term rule for the growth of bacteria is the previous term cubed. Estimate the number of bacteria in the first five days.

$t_1 = 2$

$t_2 = 2^3 = 8$

$t_3 = 8^3 = 512$

$t_4 = 512^3 = 134\,217\,728$

$t_5 = 134\,217\,728^3 = 2.419 \times 10^{24}$

Exercise 21.1

1 In each of the following sequences, let the first term (t_1) be 3.
Generate the first five terms using the term-to-term rule in each case.

   a   Double the previous term and add 1

   b   Treble the previous term and subtract 4

   c   Square the previous term and subtract 5

   d   Cube the previous term and subtract 24

   e   Double the previous term and divide by 3

LET'S TALK

Are the numbers in the answer to part (d) still considered to be a sequence?

2 In each of the following, a term in a sequence of numbers is given, as is the term-to-term rule. Work out the missing terms.

a

The term-to-term rule is double the previous term and subtract 3.

b

The term-to-term rule is treble the previous term and add 2.

c

The term-to-term rule is square the previous term.

d

The term-to-term rule is subtract 8 from the previous term and then square.

LET'S TALK

Is there more than one possible solution for the 1st term in parts (c) and (d)? Give a **convincing** answer.

3 a Below is a sequence of five numbers. Some of the terms are missing.

_____   a + 5   _____   _____   18

The term-to-term rule is double the previous term and subtract 2. Calculate:
  i) the value of a
  ii) the five terms of the sequence.

b Below is a different sequence of four numbers.

_____   2a + b   3b + 19   18a − 28

The term-to-term rule is treble the previous term and subtract 5. Calculate:
  i) the values of a and b
  ii) the four terms of the sequence.

The *n*th term

Linear sequences

Finding a rule for the *n*th term is generally more efficient if you are, say, given the first five terms of a sequence and need to find the 100th term.

This is saying the position number is multiplied by 3 and then has 2 added to give the term number, i.e. $3 \times 2 + 2 = 8$ or $3 \times 3 + 2 = 11$ or $3 \times 4 + 2 = 14$ etc.

Worked example

Consider the sequence below, in which five terms and their position within the sequence are given.

Position	1	2	3	4	5	n
Term	5	8	11	14	17	

a Calculate the term-to-term rule.

The term-to-term rule is +3.

b Write the rule for the nth term.

The rule for the nth term is $3n + 2$.

c Calculate the 100th term (t_{100}).

Substituting $n = 100$ into the rule for the nth term gives:

$3 \times 100 + 2 = 302$

Therefore $t_{100} = 302$.

KEY INFORMATION

A linear sequence is one in which the rule for the nth term takes the form $an \pm b$, e.g. $2n + 4$, $5n - 3$ or $-n + 1$.

LET'S TALK

Work through Exercise 21.2 collaboratively, discussing your ideas and findings.

The following exercise will give you an opportunity to explore the relationship between the term-to-term rule and the rule for the nth term of a **linear sequence**.

Exercise 21.2

1 Consider each of the sequences in parts (a)–(c). In each case, the terms and their position within the sequence are given.

a

Position	1	2	3	4	5	n
Term	3	5	7	9	11	

i) Calculate the term-to-term rule.
ii) Write the rule for the nth term.
iii) Calculate the 100th term.

b

Position	1	2	3	4	5	n
Term	2	5	8	11	14	

i) Calculate the term-to-term rule.
ii) Write the rule for the nth term.
iii) Calculate the 100th term.

The nth term of a sequence is sometimes called the **general term**.

When you find the nth term you are **generalising**.

c

Position	1	2	3	4	5	n
Term	6	10	14	18	22	

i) Calculate the term-to-term rule.
ii) Write the rule for the nth term.
iii) Calculate the 100th term.

d i) Enter your results from parts (a)–(c) in a table like the one below.

Sequence	Term-to-term rule	Rule for nth term
(a)		
(b)		
(c)		

ii) Write down any similarities you notice between the term-to-term rule and the rule for the nth term.

e Using your findings in part (d)(ii), work out the rule for the nth term in each of the following linear sequences. Justify your method.
i) 3, 8, 13, 18, 23
ii) −3, −7, −11, −15, −19

KEY INFORMATION

The **coefficient** means the number multiplying the variable n.

You will have seen in the previous exercise that there is a link between the term-to-term rule and the rule for the nth term in a linear sequence. The term-to-term value is the same as the **coefficient** of n.

This can be proved.

Using the sequence in the previous worked example, the rule for the nth term was $3n + 2$, which can be proved as follows:

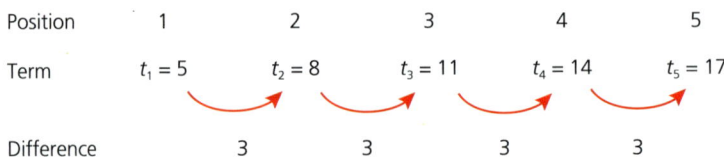

Position	1	2	3	4	5
Term	$t_1 = 5$	$t_2 = 8$	$t_3 = 11$	$t_4 = 14$	$t_5 = 17$
Difference		3	3	3	3

Notice how the term number relates to the number multiplying the term-to-term value.

To reach t_3 from t_1 involves the calculation $5 + 3 + 3$ or $5 + 2 \times 3$.

To reach t_4 from t_1 involves the calculation $5 + 3 + 3 + 3$ or $5 + 3 \times 3$.

To reach t_6 from t_1 would involve the calculation $5 + 3 + 3 + 3 + 3 + 3$ or $5 + 5 \times 3$.

To reach t_n from t_1 would involve the calculation $5 + (n − 1) \times 3$.

Therefore $\quad t_n = 5 + 3(n - 1)$

$\quad\quad\quad\quad t_n = 5 + 3n - 3 \quad$ *Expanding the brackets.*

$\quad\quad\quad\quad t_n = 3n + 2 \quad\quad$ *Simplifying the right-hand side of the equation.*

The rule for the nth term has therefore been proved to be $3n + 2$.

If d is the term-to-term value (also known as the **common difference**) then the formula for the nth term can be **generalised** as $t_n = t_1 + (n - 1)d$.

Non-linear sequences

Consider the sequence 1, 4, 9, 16, 25.

You may already have noticed that this is the sequence of square numbers. The rule for the nth term can be seen more clearly in the table below.

Position	1	2	3	4	5	n
Term	1	4	9	16	25	

Each term is the square of the position number, therefore the nth term is n^2, i.e. $t_n = n^2$.

With a linear sequence, the difference between successive terms was constant. This is not the case for **quadratic sequences**.

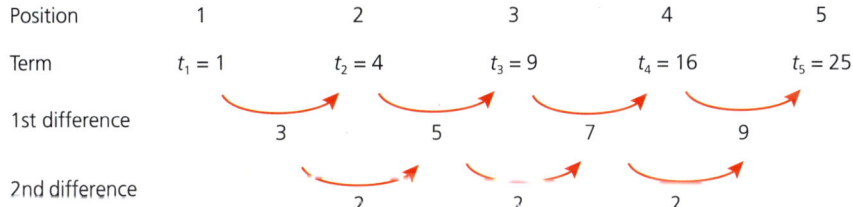

Although the first row of differences is not constant the second row of differences is.

The second row of differences will be constant for any quadratic sequence.

LET'S TALK

If the 1st row of differences is constant for a linear sequence and the 2nd row of differences is constant for a quadratic sequence, what might be the case for a cubic sequence?

How can you test your theory?

The relationship between the 2nd difference and the coefficient of n^2 and the reason for it is unfortunately beyond the scope of this book.

But, of course, you can investigate it in your own time.

Worked examples

1   The table below shows a sequence of numbers and their position in the sequence.

Position	1	2	3	4	5
Term	−4	−1	4	11	20

a   Show that the rule for the nth term is $n^2 - 5$.

   If the rule is correct, then substituting each of the position numbers into the rule will produce the term.

$$n = 1 \implies 1^2 - 5 = -4 \quad ✔$$
$$n = 2 \implies 2^2 - 5 = -1 \quad ✔$$
$$n = 3 \implies 3^2 - 5 = 4 \quad ✔$$
$$n = 4 \implies 4^2 - 5 = 11 \quad ✔$$
$$n = 5 \implies 5^2 - 5 = 20 \quad ✔$$

   Therefore $n^2 - 5$ is correct.

b   Calculate the 10th term.

$$n = 10 \implies 10^2 - 5 = 95$$

   Therefore the 10th term is 95.

c   What is the position of the term with a value of 220?

   The rule for the nth term can be rearranged to find n.

$$n^2 - 5 = 220$$
$$n^2 = 225 \qquad \text{Add 5 to both sides of the equation.}$$
$$n = \sqrt{225} = 15 \qquad \text{Square root both sides.}$$

   Therefore 220 is the 15th term.

2 The patterns of squares below follow a rule.

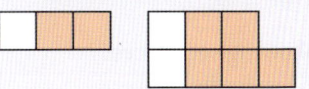 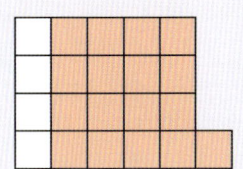

a Complete a table of results for the numbers of white and red squares in each pattern.

Number of white squares	1	2	3	4
Number of red squares	2	5	10	17

b Is the sequence of red squares linear or quadratic? Justify your answer.

The type of sequence can be deduced by looking at the differences between the terms.

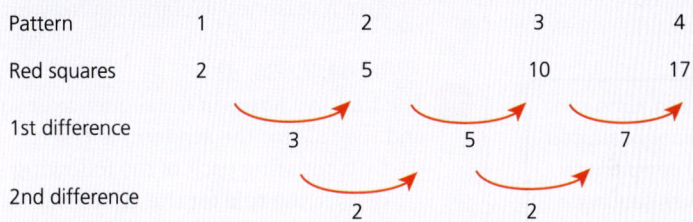

| Pattern | 1 | 2 | 3 | 4 |

As the second difference is constant, the numbers of red squares form a quadratic sequence.

c Deduce the rule for the nth term of the number of red squares.

$n^2 + 1$

d Explain how the rule for the nth term of the number of red squares relates to the diagrams.

Each pattern of red squares is formed of a square part (n^2) and then a single square added on the side (+ 1).
Therefore, the rule is $n^2 + 1$.

e Use the rule to work out the number of red squares for a pattern with 10 white squares.

The position number is equal to the number of white squares. Therefore, the rule $n^2 + 1$ can be used where n represents the number of white squares.

$n = 10 \implies 10^2 + 1 = 101$

Exercise 21.3

 1 For each of the following sequences:
   i) write the next two terms
   ii) justify your answers.
   a $-2, -6, -10, -14$
   b $3, 6, 11, 18$
   c $1, 8, 27, 64$
   d $1, 3, 6, 10, 15$
   e $\dfrac{1}{3}, \dfrac{2}{3}, 1, \dfrac{4}{3}, \dfrac{5}{3}$

2 The rules for the nth term of some sequences are given below.
In each case write down the first four terms of the sequence.
   a $t_n = n^2 + 6$
   b $t_n = \dfrac{1}{4}n + 2$
   c $t_n = -\dfrac{3}{2}n - 1$
   d $t_n = n^2 - 8$
   e $t_n = 2 - \dfrac{n}{2}$
   f $t_n = n^3$

 3 For each of the following sequences:
   i) decide whether it is linear, quadratic or neither
   ii) give a reasoned explanation for your answer to part (i).
   a $-3, 3, 13, 27$
   b $14, 20, 26, 32$
   c $-3, 4, 23, 60, 121$
   d $4, 1, -4, -11, -20$

4 The first five terms in the sequence of square numbers are 1, 4, 9, 16, 25
and the rule for the nth term is $t_n = n^2$.
   i) By comparing each of the following sequences to the one above,
   deduce the rule for the nth term.
   ii) Write down the 10th term.
   a $4, 7, 12, 19, 28$
   b $-1, 2, 7, 14, 23$
   c $7, 10, 15, 22, 31$
   d $-1, -4, -9, -16, -25$

> Often, rules can be easily deduced by comparing a given sequence to another well-known one.

 5 The patterns of blue and white squares below form a sequence.

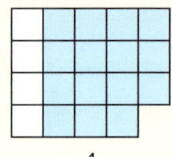

1   2   3   4

   a Copy and complete the table below for the number of blue squares
   in each pattern.

Position	1	2	3	4
Number of blue squares				

   b Write down the rule for the nth term of the number of blue squares
   in the sequence.
   c Explain, with reference to the patterns themselves, why the rule works.
   d Write down the numbers of white and blue squares in the 14th
   pattern.

6 The pattern sequence below shows white and orange tiles.
The area of a square tile is 1 unit^2.

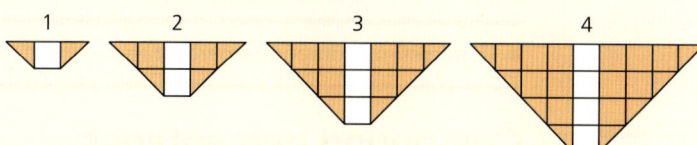

1   2   3   4

 a Write down the rule for the nth term of the area of the orange tiles.
 b Explain, with reference to the patterns, why the rule for the nth term works.
 c **i)** How many square orange tiles will be needed for the 10th pattern?
   ii) How many triangular orange tiles will be needed for the 10th pattern?

 7 A sequence of cuboids made from small cubes is shown below.

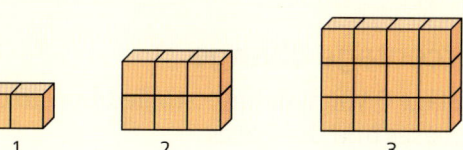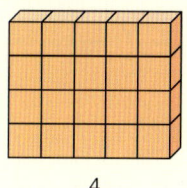

1   2   3   4

The number of rows and the number of columns of cubes increase by 1 each time.
 a If n is the position number, explain why the formula for the nth term of the number of cubes can be given by $t_n = n(n+1)$.
 b How many cubes will be needed for the 10th cuboid?

Now you have completed Unit 21, you may like to try the Unit 21 online knowledge test if you are using the Boost eBook.

22 Compound percentages

- Understand compound percentages.

LET'S TALK

What examples can you think of where people borrow money and pay it back over long periods of time?

Compound percentages

Compound percentages are used widely in the world of finance. Usually, whenever money is borrowed or lent over a period of time, compound percentages in the form of interest are used.

In many cases, the use of compound percentages when borrowing money, means that the borrower ends up paying back a lot more over time than the original amount borrowed.

To understand why, an understanding of percentage increases is needed.

Percentage change

In Stage 8 you will have studied how to calculate percentage increases and decreases.
For example, if a quantity is increased by 22% this can be visualised as follows.

Imagine the original amount being 100%.

A 22% increase therefore means that we now have 122% compared to the original amount:

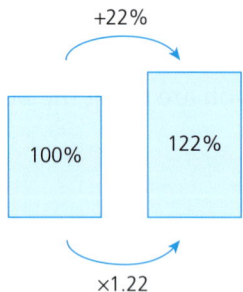

The diagram shows that an increase in 22% can be done in two ways:
1 by adding 22%
2 by multiplying by 1.22 because 100 × 1.22 = 122.

The second method is particularly useful if we need to work out the actual amount after a percentage increase.

> Although the multiplication came to 146.4, as the question is dealing with money, the answer should include the units and be given to 2 d.p., i.e. $146.40.

Worked example

The price of a coat in a sale is $120.
When the sale ends, the price of the coat is increased by 22%.

Calculate the new price of the coat.

A 22% increase is equivalent to multiplying by 1.22.

Therefore, this is $120 \times 1.22 = 146.4$.

The new price of the coat is $146.40.

With a percentage decrease the same method can be applied.
For example, if a quantity is decreased by 15% this can be visualised as follows.

Imagine the original amount being 100%.

A 15% decrease therefore means that we have now 85% compared to the original amount:

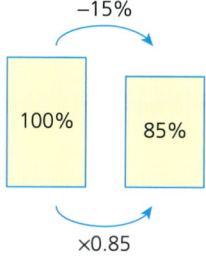

The diagram shows that a decrease of 15% can be done in two ways:
1 by subtracting 15%
2 by multiplying by 0.85 because $100 \times 0.85 = 85$.

Once again, the second method is particularly useful if we need to work out the actual amount after a percentage decrease.

Worked example

The cost of a car when new is $12500. After 1 year its value has decreased by 15%.

What is the value of the car after 1 year?

A 15% decrease is equivalent to multiplying by 0.85.

Therefore, this is $12500 \times 0.85 = 10625$.

The value of the car after 1 year is $10625.

Compound interest

Many adults are lent money by banks or from shops to buy large items such as houses, cars or large electrical goods. The money is lent with interest and the person borrowing the money will eventually not only have to pay back the amount borrowed, but also the interest money too.

Imagine a person borrowing $15 000 in order to buy a car. Assume the interest rate is 5% compounded per year.

Assuming that nothing is paid back in the first three years, the amount owed after three years can be calculated as follows:

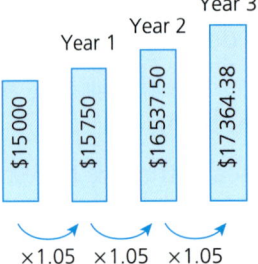

Therefore, if nothing has been paid back for three years, the amount owed at the end is $17 364.38. The compounded interest added on over the three years is $2364.38.

Rather than drawing a diagram each time, the total can be worked out using the following calculation:

$$15 000 \times 1.05 \times 1.05 \times 1.05 = 17 364.38$$

The calculation is more efficiently written using indices, as:

$$15 000 \times 1.05^3 = 17 364.38$$

Worked examples

1   Carla deposits $5000 in a savings account. The interest rate is 3% compounded each year. Assuming she does not take any money out and that the interest rate remains constant, how much will she have in her account after 10 years?

3% is equivalent to a multiplier of $\times 1.03$.

$$5000 \times 1.03^{10} = 6719.58$$

Therefore, Carla has $6719.58 in her account after 10 years.

2   A computer costs $2500 when new.

Its value is expected to depreciate by 12% compounded per year.

How much will the computer be worth after 5 years?

A 12% decrease is equivalent to a multiplier of $\times 0.88$.

$$2500 \times 0.88^5 = 1319.33$$

Therefore, the computer will have a value of $1319.33 after 5 years.

Exercise 22.1

 1 On her 10th birthday a child is given $10 to put into a savings account. The interest rate is 4% compounded per year. The child only takes the money out on her 18th birthday.
Below are some calculations.
$$10 \times 4 \times 8 \qquad 10 \times 1.4^8 \qquad 10 \times 1.4 \times 8 \qquad 10 \times 1.04^8 \qquad 10 \times 1.04 \times 8$$
Which is the correct one for calculating the amount in the account on her 18th birthday?

 2 **a** **i)** Which of the following multipliers is equivalent to a 7% increase?
$$\times 1.07 \qquad \times 7 \qquad \times 1.7 \qquad \times 107$$
ii) Justify your choice.
b **i)** Which of the following multipliers is equivalent to a 62% decrease?
$$\times 1.62 \qquad \times 0.62 \qquad \times (-0.62) \qquad \times 0.38$$
ii) Justify your choice.

 **3** The price of a pair of shoes is normally $95. In the sale the price is reduced by 30%.
a What is the sale price?
b After the sale the price is increased by 30% again.
i) What are the shoes' new non-sale price?
ii) Explain why reducing something by 30% and then increasing it again by 30% will not give the original amount.

4 As part of a promotion, a mobile phone supplier reduces the cost of a particular model by 40%.
Once the promotion is over, what must the percentage increase in the price of the phone be, to return it to its original price?

 5 It is estimated that the population of a colony of rats will increase by 3% each day. If the colony has 20 rats at the start of the study, calculate how many days it will take for the colony to:
a double in size **b** treble in size.

 6 Write down a possible scenario that would require each of the following calculations.
a 6×1.55^8 **b** 400×0.56^{12}

7 Two friends, Zack and Freda, decide to invest some money in a savings account.
Zack deposits $3000 in an account offering a 3.5% compound interest rate per year.
Freda deposits $4200 in an account offering a 1.8% compound interest rate per year.
a Assuming that neither of them takes any money out and that the interest rates stay constant, who will have $6000 in their account first?
b How long did it take?

8 Banks often give loans to people and charge interest compounded each year. However, it does not always have to be compounded yearly. A yearly interest rate could be halved and compounded every six months.

As an example, a $500 loan with an 8% annual interest rate could be compounded every 6 months, and the amount owed at the end of the year would be 500×1.04^2.

a $500 is borrowed with an annual interest rate of 12%.
   What is the amount owed at the end of the year if:
   i)   it is compounded at the end of the 12 months?
   ii) it is compounded every 6 months?
   iii) it is compounded every 3 months?
   iv) it is compounded every month?

b $100 000 is borrowed with an annual interest rate of 6% for 10 years. How much more is paid if the repayment is compounded every month for the 10 years, compared with an annually compounded interest payment?

LET'S TALK

Why might a lender prefer to compound the interest calculations on a more regular basis rather than just once a year?

Now you have completed Unit 22, you may like to try the Unit 22 online knowledge test if you are using the Boost eBook.

23

Scale and area factors of enlargement

● Analyse and describe changes in perimeter and area of squares and rectangles when side lengths are enlarged by a positive integer scale factor.

Scale and area factors of enlargement

You will already be aware, from the work done on enlargement in Unit 13, that if an object undergoes an enlargement by a particular scale factor, then all of its lengths will be multiplied by that scale factor, e.g.

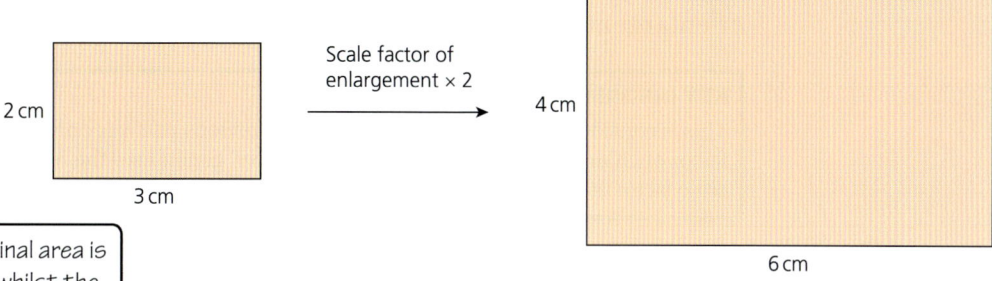

The original area is 6 cm², whilst the enlarged rectangle has an area of 24 cm².

It can be seen that although all the lengths have been doubled, the area of the original shape has more than doubled. The area has in fact increased by a factor of 4.

The reason why enlarging by a scale factor of 2 causes the area to increase by a factor of 4 can be proved as follows:

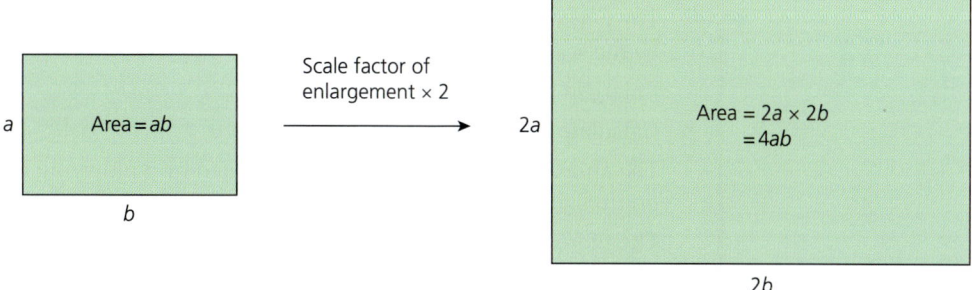

The way in which the area factor is related to the scale factor can be **generalised** with the following proof:

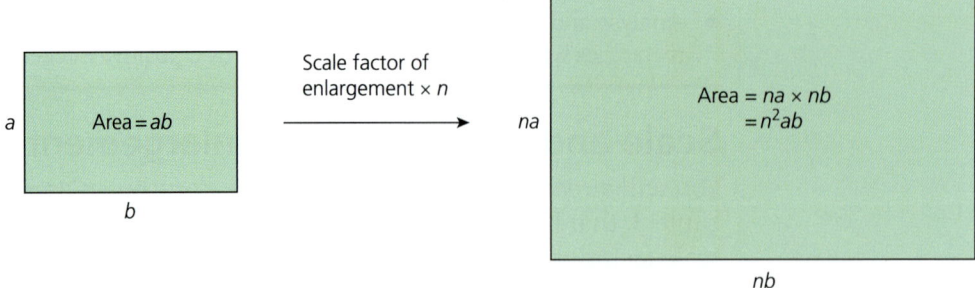

Therefore, if the scale factor of enlargement is n, then the **area factor of enlargement** will be n^2.

> **KEY INFORMATION**
>
> This is considered to be a **general** proof. The previous one was only for a scale factor of enlargement of $\times 2$. This proof, however, generalises the result for any scale factor of enlargement n.

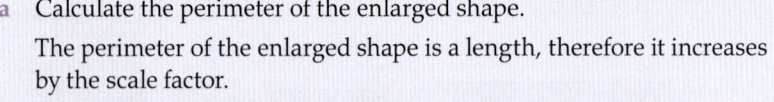

Worked example

An irregular shape is shown.

Its perimeter is 68 cm and its area is 155 cm^2.

The shape is enlarged by a scale factor of 3.

a   Calculate the perimeter of the enlarged shape.

   The perimeter of the enlarged shape is a length, therefore it increases by the scale factor.

   Enlarged perimeter = 68 × 3 = 204 cm.

b   Calculate the area of the enlarged shape.

   The area of the enlarged shape is calculated by multiplying the original area by the square of the scale factor.

   Therefore, enlarged area = 155 × 3^2 = 1395 cm^2.

Exercise 23.1

1 Below are a set of scale factors of enlargement and a set of area factors of enlargement.

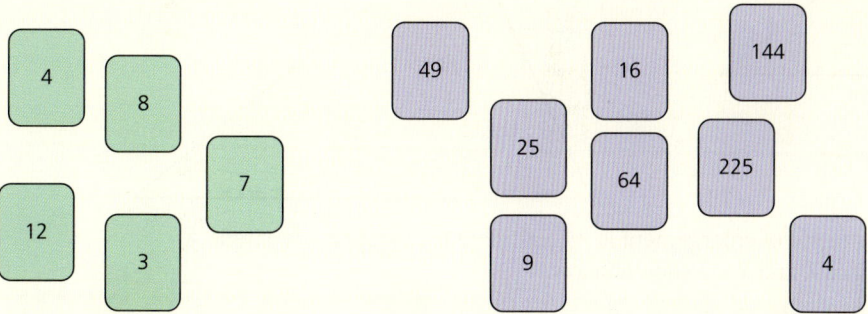

Scale factors of enlargement   Area factors of enlargement

  a Match the scale factors of enlargement with their corresponding area factors of enlargement.

  b For the unmatched area factors, work out their corresponding scale factors.

2 Two rectangles and some dimensions are shown.

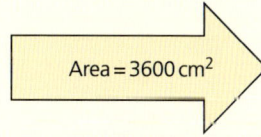

One rectangle is the enlargement of the other.
Calculate the area of an enlarged rectangle.

3 Two arrow shapes are shown. One is an enlargement of the other.

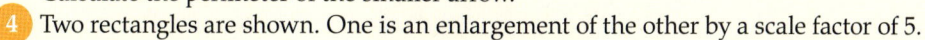

Calculate the perimeter of the smaller arrow.

4 Two rectangles are shown. One is an enlargement of the other by a scale factor of 5.

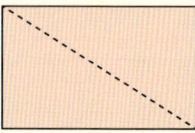

  a Calculate the area of the enlarged rectangle.
  b Calculate the length of the diagonal of the enlarged rectangle.

5 Two triangles A and B are shown. B is an enlargement of A.

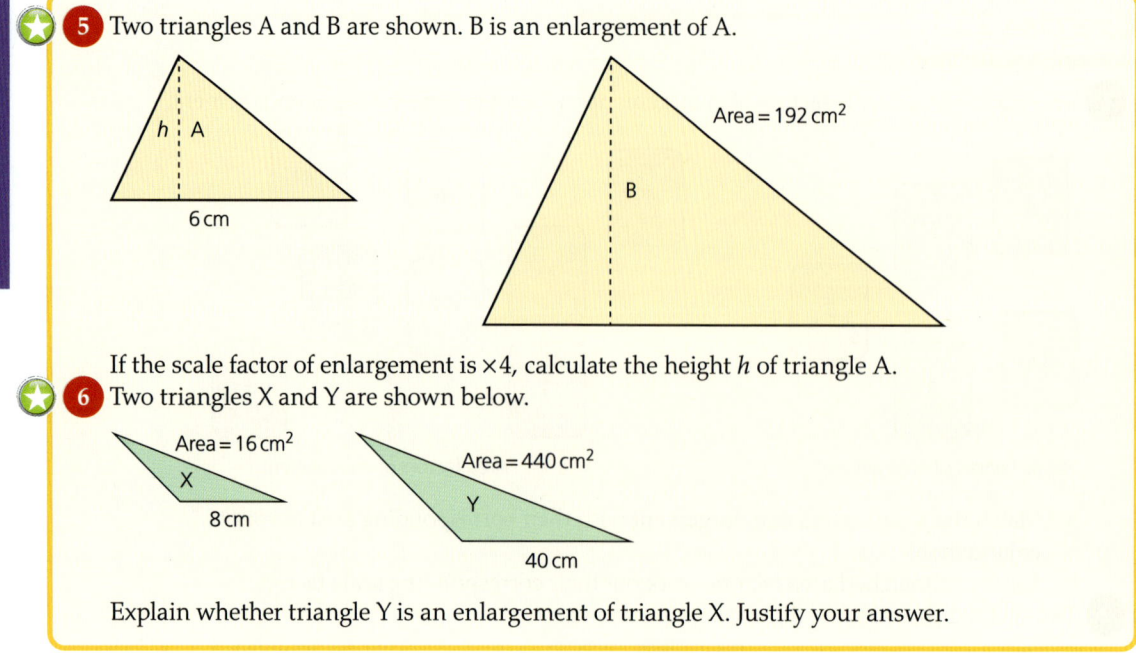

Area = 192 cm²

If the scale factor of enlargement is ×4, calculate the height h of triangle A.

6 Two triangles X and Y are shown below.

Area = 16 cm²

X

8 cm

Area = 440 cm²

Y

40 cm

Explain whether triangle Y is an enlargement of triangle X. Justify your answer.

Now you have completed Unit 23, you may like to try the Unit 23 online knowledge test if you are using the Boost eBook.

24

Functions and their representation

- Understand that a function is a relationship where each input has a single output. Generate outputs from a given function and identify inputs from a given output by considering inverse operations (including indices).
- Understand that a situation can be represented either in words or as a linear function in two variables (of the form $y = mx + c$ or $ax + by = c$), and move between the two representations.

Inputs and outputs

Recap

You will have seen that functions (or the relationship between two variables) can be presented in different forms. These include:

Equations

For example, $y = 4x + 2$

Function machines

Where the x is the input and the y is the output.

Inverse function machines

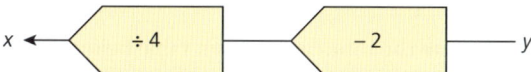

Mapping diagrams

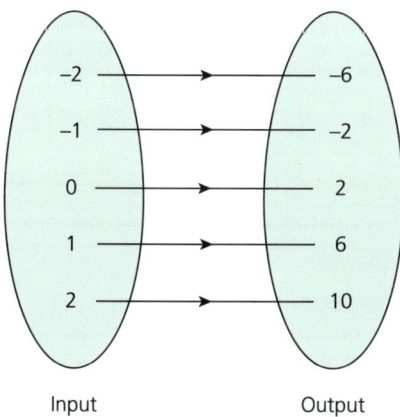

Input   Output

Graphs

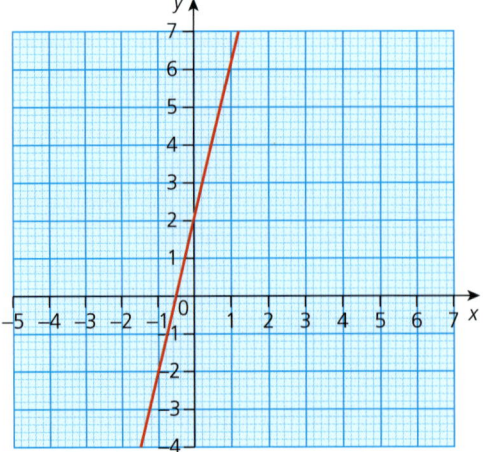

In each of the cases, one input value produces only one output value. This is a necessary condition for a relationship between two variables to be classed as a function.

Worked example

a Write a function machine for $y = 2x^2$.

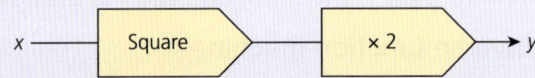

b Complete the input and output table below.

Input	Output
−2	
0	
$\frac{1}{2}$	
3	

$\Longrightarrow$

Input	Output
−2	8
0	0
$\frac{1}{2}$	$\frac{1}{2}$
3	18

c If the output is 2, what are the possible values for the input? Justify your answer.

±1 because $1^2 = (−1)^2$

Exercise 24.1

1 For each of the following, copy and complete the table of results.
If there are two possible answers, include both of them.

   a $y = x^2 + 10$ **b** $y = 3x^2$ **c** $y = -x^2$

Input	Output
−3	
0	
$\frac{1}{3}$	
4	

Input	Output
−5	
0	
$\frac{1}{2}$	
	12

Input	Output
−2	
$-\frac{1}{2}$	
$\frac{1}{2}$	
	−25

2 Two graphs A and B are given below.

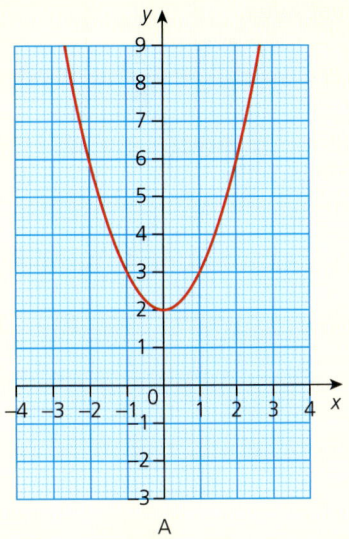

A

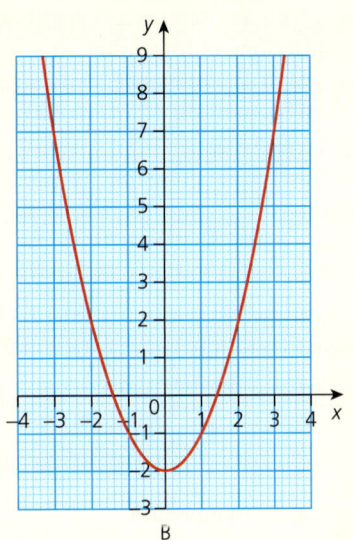

B

  a **i)** Match each of the mapping diagrams P and Q below to the correct graph.

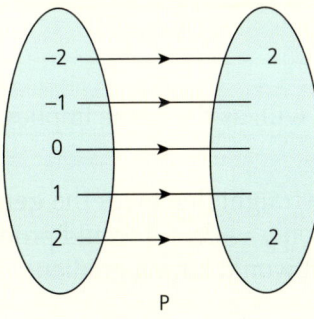

P

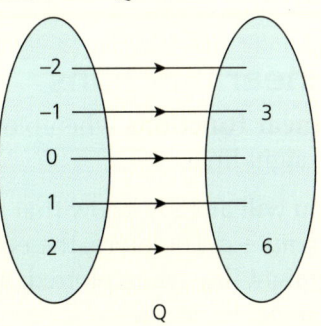

Q

      ii) Copy and complete each of the mapping diagrams.

b i) Match each of the function machines M and N below to the correct mapping diagram and the correct graph.

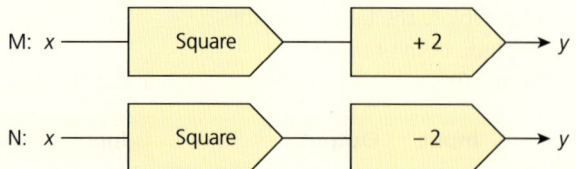

ii) Write each function machine as an equation involving y and x.

iii) If the output is −1.5, give a reasoned explanation for which function machine must have produced it.

3 a Draw a function machine for the equation $y = x^3$.

b Copy and complete the table of results.

c Draw the inverse function machine for the equation $y = x^3$.

d i) Sketch a graph of the function.

ii) Using your graph as a reference, explain why one input value always produces one output value.

iii) Using your graph as a reference, explain why one output value can only have been produced from one input value.

Input	Output
−1	
0	
$\frac{1}{2}$	
2	

4 An equation is given as $y = 5x^2$.

A student decides that the function machine for this equation is:

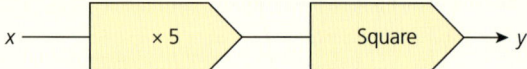

a Explain why this function machine is incorrect for $y = 5x^2$.

b What is the equation for the above function machine?

5 A function machine is given as:

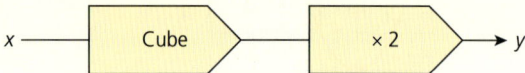

a Write the function as an equation.

b A student gets an output of 250 and states that the input can have been ±5. Is the answer correct? Justify your answer.

Linear functions

KEY INFORMATION

m represents the gradient of the straight line and c the intercept with the y-axis.

Linear functions when plotted will, as the name implies, produce a straight line.

You will already know that any straight line takes the **general** form $y = mx + c$. Therefore, the equation $y = -3x + 4$ would produce a straight line when plotted, and it would have a gradient of −3 and a y-intercept of +4.

LET'S TALK

What advantages are there of a linear equation given in the form $y = mx + c$?

What advantages might there be for a linear equation to be of the form $ax + by = c$?

The equation could be rearranged, though, so that the x and y terms are on the same side of the equation, i.e. $y = -3x + 4$ can be written as $3x + y = 4$ by adding $3x$ to both sides of the equation, and both would produce the same straight line if plotted.

This format is known as writing a linear equation in the form $ax + by = c$.

The values of a, b and c are usually integer values.

Plotting and manipulating equations in this format will be covered in more detail in Unit 27 of this book.

LET'S TALK

Rearranging in a different order produces this:

$y - 4x = -3$
   Subtract 4x.

$-4x + y = -3$
   Reorder.

Is $-4x + y = -3$ the same equation as $4x - y = 3$?

The equations $\frac{2}{3}x - y = -2$ and $2x - 3y = -6$ are different ways of writing the same linear relationship between x and y. If plotted, they would produce the same straight line.

Worked examples

1 Write each of the following linear equations in the form $ax + by = c$.

   a $y = 4x - 3$

      $0 = 4x - y - 3$ *Subtract y from both sides.*

      $3 = 4x - y$ *Add 3 to both sides.*

      $4x - y = 3$ *Write in the required order.*

   b $y = \frac{2}{3}x + 2$

      $0 = \frac{2}{3}x - y + 2$ *Subtract y from both sides.*

      $-2 = \frac{2}{3}x - y$ *Subtract 2 from both sides.*

      $\frac{2}{3}x - y = -2$ *Write in the required order.*

   In order to eliminate the fractional part of the equation, the whole equation can be multiplied by 3.

      $2x - 3y = -6$ *Multiply each term by 3.*

LET'S TALK

What other ways could these equations be written?

2 A farmer keeps geese and sheep.

   a If there are 120 legs in total, write an equation linking the number of geese (g), the number of sheep (s) and the total number of legs.

   As each goose can be assumed to have 2 legs, then $2g$ is the expression for the total number of goose legs.
   As each sheep can be assumed to have 4 legs, then $4s$ is the expression for the total number of sheep legs.
   Therefore, the equation is $2g + 4s = 120$.

b   Give one possible combination of geese and sheep that would fit this equation.

As there are two variables and only one equation, there are likely to be many possible solutions. Therefore a strategy is to fix one of the variables.

Let $g = 2$, i.e. the farmer has two geese.

Substitute this into the equation and then solve it:

$$2(2) + 4s = 120$$
$$4 + 4s = 120$$
$$4s = 116$$
$$s = 29$$

LET'S TALK

Can g be any whole number or are there restrictions?

Therefore, one possible combination is that the farmer has 2 geese and 29 sheep.

Exercise 24.2

 1 A student is trying to rearrange the equation $y = 5x + 1$ into the form $ax + by = c$.

He does the following rearranging:
$$y = 5x + 1$$
$$y - 5x = 1$$
$$-5x + y = 1$$

Another student rearranges it a different way:
$$y = 5x + 1$$
$$0 = 5x - y + 1$$

a   Are both methods correct?
b   Justify your answer.

 2 Three linear equations are given below in different formats.
Classify the different formats for each equation into groups.

$y = 4x + 8$	$4x + y + 8 = 0$	$-4x + y = 8$
$-x + 4y = -8$	$4x + y = -8$	$y - 4x = 8$
$y = -4x - 8$	$-4x + y - 8 = 0$	$y = \frac{1}{4}x - 2$
$y + 4x = -8$	$4y = x - 8$	$4y - x = -8$

3 A furniture store sells three- and four-legged stools.
In the store there are a total of 632 stool legs.
  a Write an equation linking the number of three-legged stools (t),
    the number of four-legged stools (f) and the total number of legs.
  b If the number of three-legged stools is greater than 50, work out a
    possible combination of three- and four-legged stools in the store.
  c i) Explain why there cannot be 220 three-legged stools.
    ii) Explain why there cannot be 65 three-legged stools.

4 A restaurant has large tables that can seat 8 people and smaller tables
  that can seat 4 people. The restaurant has a capacity for 148 people.
  a Write an equation linking the number of large tables (L), the
    number of small tables (S) and the maximum capacity in the form:
    i) $aL + bS = c$
    ii) $S = mL + c$
  b Give a **convincing** explanation why there cannot be 18 small
    tables in the restaurant.
  c i) If there are more large tables than small tables in the
       restaurant, give a possible combination of large and small
       tables in the restaurant.
    ii) Justify your answer to part (i).

5 A café sells cups of coffee for $1.50 each and cups of tea for $1 each.
  One afternoon, the café works out that it has received a total of $77
  from selling teas and coffees.
  a Write an equation linking the number of coffees sold (c), the
    number of teas sold (t) and the total amount received.
  b If 20 coffees were sold, how many teas were sold?
  c i) What is the minimum number of teas that can have been sold?
    ii) Justify your answer to part (i).
  d If 3 times as many coffees were sold compared to teas, calculate
    the number of teas sold.
  e Explain why there cannot have been twice as many coffees sold
    compared to teas.

Note: Part
(a)(i) is asking you
to write it in the
form $ax + by = c$,
whilst part (a)(ii)
is asking for it to
be written in the
form $y = mx + c$.

Now you have completed Unit 24, you may like to try the Unit 24
online knowledge test if you are using the Boost eBook.

Coordinates and straight-line segments

- Use knowledge of coordinates to find points on a line segment.

Line segments

In mathematics a **straight-line segment** has a very specific meaning. It is a straight line that joins two points. What we usually draw in our books or on paper and call a line are in fact straight-line segments, because a true line goes on for ever and has no beginning or end.

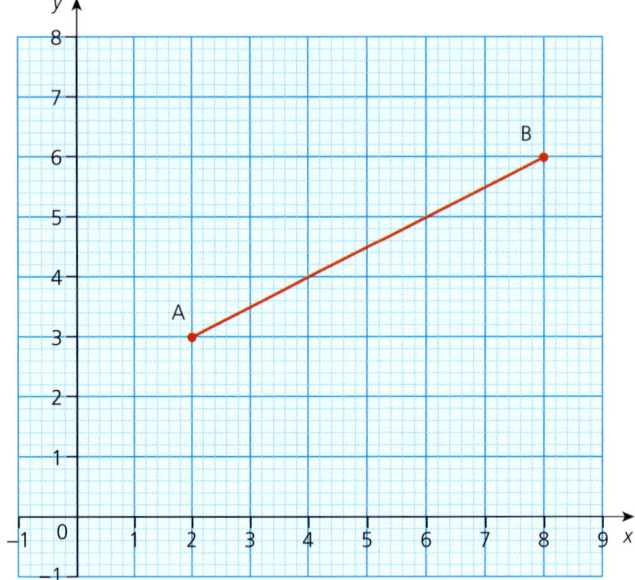

LET'S TALK
How would you calculate the length of a line segment?

In the diagram, the straight-line segment AB joins the two points A and B.

As it is a straight line, its length also represents the shortest distance between A and B.

Recap

You will have covered how to calculate the midpoint of a line segment in Stage 8.

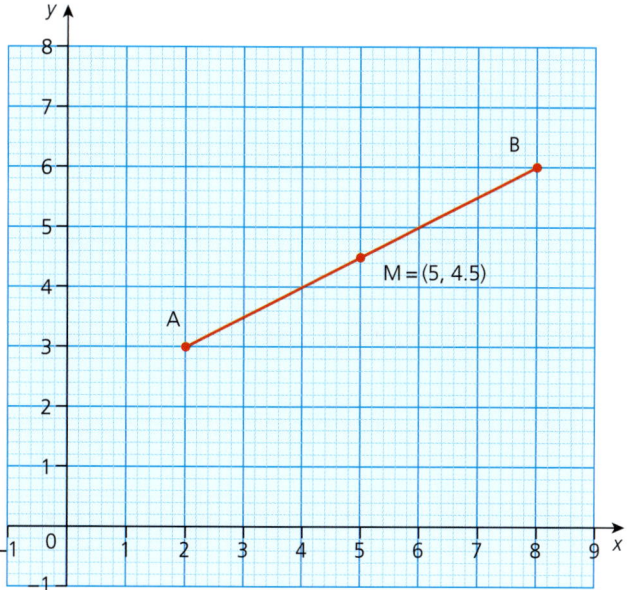

For A and B above, the midpoint can be found by calculating the mean value of their x-coordinates and the mean value of their y-coordinates. A has coordinates (2, 3) and B has coordinates (8, 6), therefore the midpoint (M) is

$$\left(\frac{2+8}{2}, \frac{3+6}{2}\right) = (5,\ 4.5)$$

Another method could be to visualise halfway between the x-coordinates and halfway between the y-coordinates, i.e. halfway between 2 and 8 is 5, whilst halfway between 3 and 6 is 4.5.

Other calculations involving straight-line segments are best answered by visualising the problem first with a good diagram.

Worked examples

1 Consider the straight-line segment XY shown.

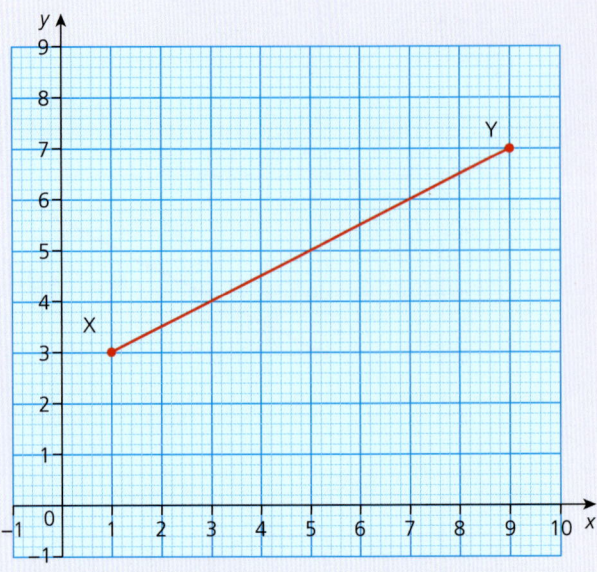

A point P, lies on the line XY such that it is three times as far from X as from Y. Calculate the coordinates of P.

If the distance XY is split into quarters, then point P would be $\frac{3}{4}$ of the way along the length as shown.

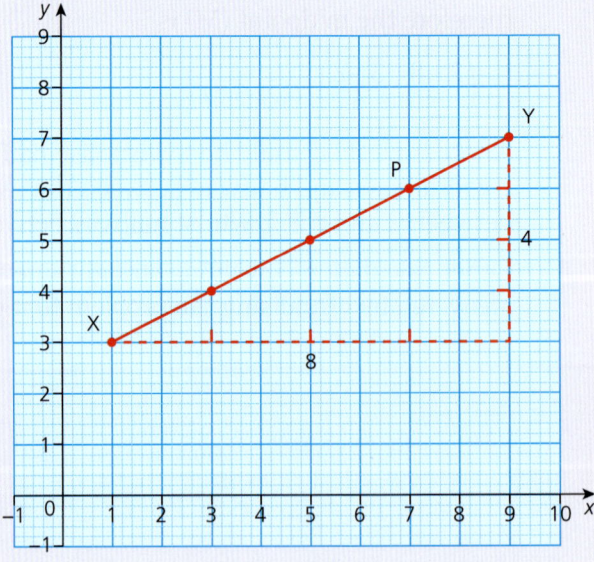

The coordinates of P can be deduced from the diagram as (7, 6).

The coordinates of P can also be calculated as follows:

The difference in the x-coordinates between X and Y is 8, and $\frac{3}{4}$ of 8 = 6.

But this must be added on to the x-coordinate of point X, i.e. 1 + 6 = 7.

The difference in the y-coordinates between X and Y is 4, and $\frac{3}{4}$ of 4 = 3.

But this must be added on to the y-coordinate of point X, i.e. 3 + 3 = 6.

Therefore, the coordinates of P are (7, 6).

2 A point Q is at the midpoint of the straight-line segment PR.

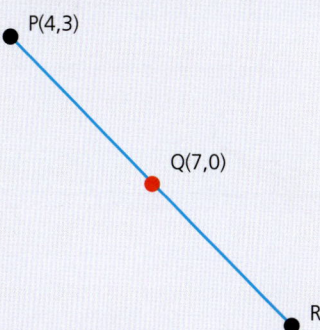

The coordinates of P are (4, 3) and the coordinates of Q are (7, 0).

Calculate the coordinates of point R.

A quick sketch of the information is shown below.

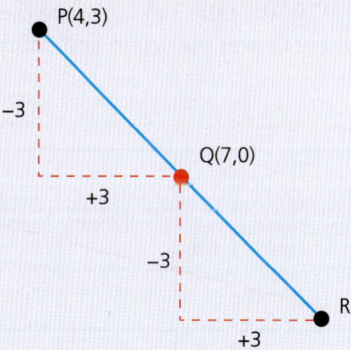

As Q is the midpoint of PR, the distance PQ = QR.

Therefore, the x-coordinate of R is 4 + 3 + 3 = 10.

The y-coordinate of R is 3 − 3 − 3 = −3.

The coordinates of R are (10, −3).

Exercise 25.1

1 Which of the following pairs of coordinates has the point M(3,4) as its midpoint?
  a (1,2) and (5,6)   c (–1,7) and (7,1)
  b (0,0) and (6,8)   d (–2,4) and (8,4)

2 The straight-line segment OA has end coordinates O(0, 0) and A(8, 12).
Calculate the coordinates of the following points:
  a M, the midpoint of OA   d D, if A is the midpoint of OD
  b B, if B divides OA in the ratio 1:3   e E, if A is the midpoint of ME
  c C, if C divides OA in the ratio 3:1   f F, if O is the midpoint of AF.

> A simple sketch each time will help to visualise each question.

3 A kite ABCD is shown.

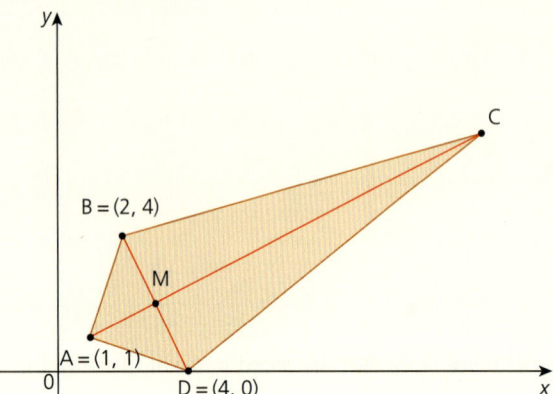

  a Calculate the coordinates of the midpoint M of BD.
  b If AM divides the line segment AC in the ratio 1:5, calculate the coordinates of point C.

4 A straight-line segment is drawn where O is at the origin and points A, B, C etc. are equally spaced along its length as shown.

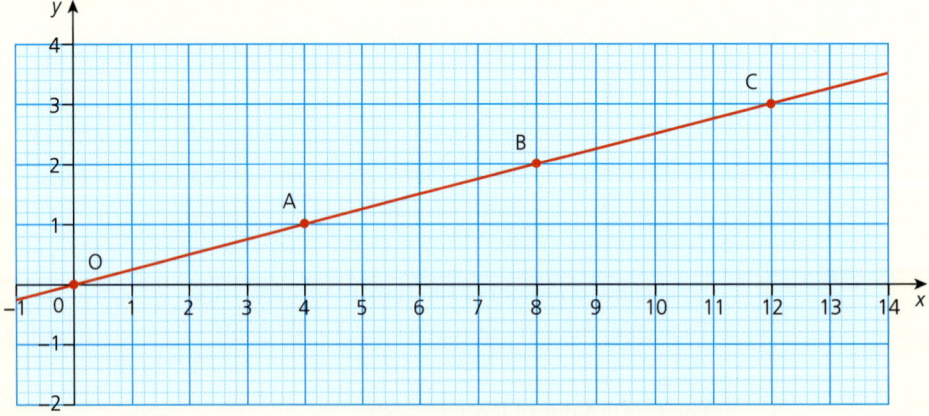

a Describe how the coordinates of the points change from one point to the next.
b Assuming the points follow the letters of the alphabet, deduce the coordinates of point J.
c i) Which letter has the coordinates (64, 16)?
   ii) Justify your answer in part (i).
d Work out the length of the straight-line segment CF.

5 A straight-line segment XY has end coordinates X(−6, 8) and Y(14, −12).
a Calculate the coordinates of M, the midpoint of XY.
b The point P divides XY in the ratio 3:2. Calculate the coordinates of P.
c The point Y divides the straight-line segment XZ in the ratio 5:2.
   Calculate the coordinates of Z.

6 A straight-line segment has a number of points equally
spaced along its length as shown.
Point A has coordinates (0, 8) and point D has
coordinates (12, 0).
a Calculate the coordinates of point C.
b Assuming the points remain evenly spaced and
   their labelling follows the sequence of the alphabet,
   calculate the coordinates of point X.
c Explain whether a labelled point will have
   coordinates (36, −14).

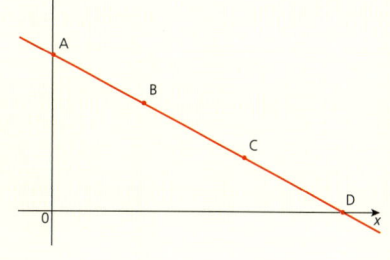

7 Five squares are stacked as shown on a pair of axes.
The bottom square has a side length of four units and its bottom
left corner is at the origin (0, 0). Each square has a side length half
the length of the square directly below it. The bottom right-hand
corner of each square is highlighted and labelled A, B, C, D etc.
a Calculate the coordinates of point C.
b Calculate the coordinates of the end points of the straight-line
   segment BE.
c Prove that the straight-line segment BE also passes through
   points C and D.
d Prove that C is not the midpoint of BE.
e i) What is the height of the five stacked squares?
   ii) If the sequence of squares and labelling is continued, what
       is the height of a stack of nine squares?
   iii) What are the coordinates of the end points of the straight-line segment HJ?

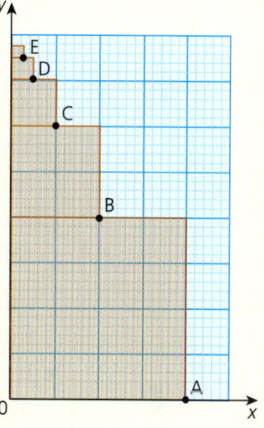

> **LET'S TALK**
>
> Assuming the square stacking continues, is there, in theory, a maximum height the stack
> of squares will reach?

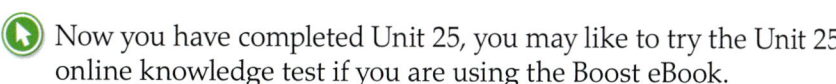

Now you have completed Unit 25, you may like to try the Unit 25
online knowledge test if you are using the Boost eBook.

- Use knowledge of square and cube roots to estimate surds.

Surds

Numbers belong to different groups. For example, you have already encountered **integers** and **rational** and **irrational numbers**.

In Unit 9 of this book you were introduced to a hierarchy within the number system, presented as shown below.

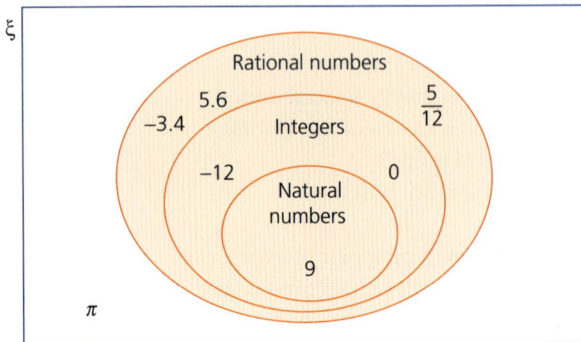

LET'S TALK

What does each of these terms mean?

The nested arrangement of the oval shapes implies that those numbers which are classed as natural numbers also belong to the family of integers. These in turn also belong to the larger family of rational numbers.

However, from the diagram it can also be seen that some numbers fall outside the 'rational' family of numbers and these are called irrational numbers. The one example of an irrational number shown above is π.

An irrational number is one where the number cannot be written as a fraction in which both numerator and denominator are integer values.

One of the features of an irrational number is that, when written as a decimal, it does not terminate or repeat itself.

Another type of irrational number is **surds**.

A surd is a square root or cube root of an integer that does not produce an integer result.

KEY INFORMATION

Although only square roots and cube roots will be covered here, a surd can be the nth root of any number that does not produce an integer result.

LET'S TALK

Why is $\sqrt{\pi}$ not considered to be a surd?

Therefore, $\sqrt{5}$ is a surd, but $\sqrt{16}$ is not as $\sqrt{16} = 4$.

Similarly, $\sqrt[3]{10}$ is a surd, but $\sqrt[3]{27}$ is not as $\sqrt[3]{27} = 3$.

Knowledge of surds is also a good way to check answers, as shown in the worked example below.

Worked example

A student is working with a square of area $70\,\text{cm}^2$.

He calculates the length of each side of the square to be $7.3\,\text{cm}$.

Show, without working out the exact answer, why 7.3 must be incorrect.

If the area $= 70\,\text{cm}^2$, then the length of each side must be $\sqrt{70}\,\text{cm}$ as $\sqrt{70} \times \sqrt{70} = 70$.

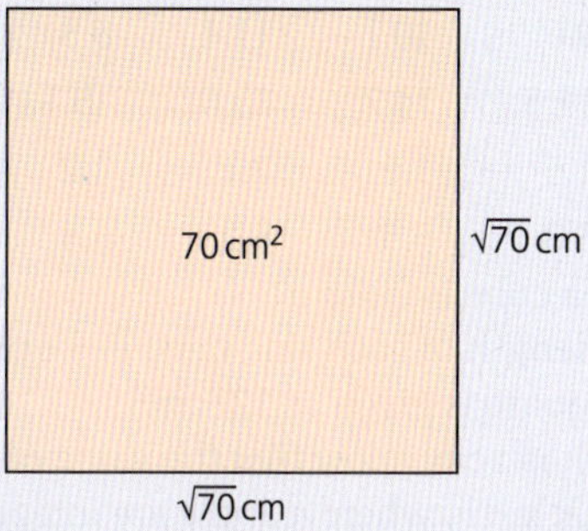

$70\,\text{cm}^2$ $\sqrt{70}\,\text{cm}$

$\sqrt{70}\,\text{cm}$

The square numbers either side of 70 are 64 and 81.

As $\sqrt{64} = 8$ and $\sqrt{81} = 9$, then $\sqrt{70}$ must lie between 8 and 9.

This can be written as an inequality as $8 < \sqrt{70} < 9$.

As 7.3 does not lie in this range it must be incorrect.

Exercise 26.1

1 **a** **Classify** these numbers into two groups: surds and not surds.

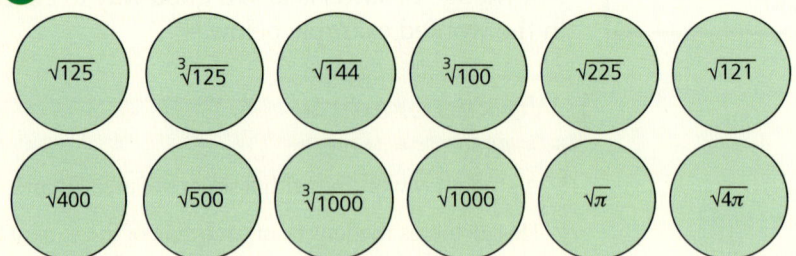

$\sqrt{125}$ $\sqrt[3]{125}$ $\sqrt{144}$ $\sqrt[3]{100}$ $\sqrt{225}$ $\sqrt{121}$

$\sqrt{400}$ $\sqrt{500}$ $\sqrt[3]{1000}$ $\sqrt{1000}$ $\sqrt{\pi}$ $\sqrt{4\pi}$

b Justify each of your choices.

2 In each of the following, identify which of the given answers are definitely incorrect.

  a $\sqrt{92} =$ 9.6   10.1   11.4   8.8

  b $\sqrt{180} =$ 14.5   14.7   13.6   15.1   13.4

  c $\sqrt[3]{4} =$ 1.5   1.6   2.1   2.2

  d $\sqrt[3]{152} =$ 5.3   5.5   6.2   6.4   7.1

3 The volume of a cube is $200\,\text{cm}^3$.
Prove that the length of each side must lie between 5 cm and 6 cm.

4 The area of a square is given as $127\,\text{cm}^2$.
If its side length is x cm, show that the value of x can be written as an inequality as $a < x < b$, where a and b are integer values.

5 A rectangle has dimensions as shown.

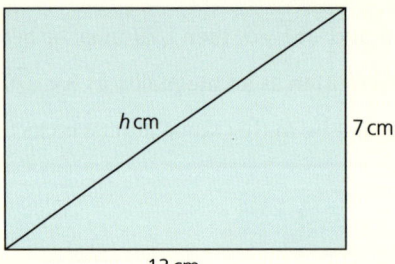

h cm

7 cm

12 cm

  a Prove that $h = \sqrt{193}$.

  b Write the value of h in the form $a < h < b$, where a and b are the nearest integer values to h.

6 **a** Prove that $\sqrt{200}$ must lie between 14 and 15.

  b Is $\sqrt{200}$ closer to 14 or 15?
Give a reasoned justification for your answer.

7 A rectangle has dimensions as shown.

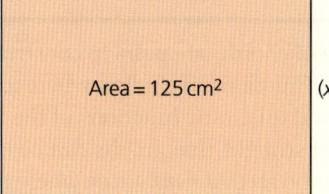

Area = 125 cm² $(x-5)$ cm

$(x+5)$ cm

   a Write the value of x in surd form.
   b Write the value of x in the form $a < x < b$, where a and b are the nearest integer values to x.

 8 The volume of a cube is 900 cm³.
   a Write the length (L) of each edge of the cube in the form $a < L < b$, where a and b are the closest integer values to L.
   b Using your results from part (a), show that the total surface area (A) of the cube must lie in the range $p < A < q$, where p and q are integers.
   c Two students, Thea and Joe, are trying to work out the total surface area of the cube.
     Thea says it's 559 cm², whilst Joe calculates it as being 605 cm².
     Using your answer to part (b), explain which one of them must be wrong.

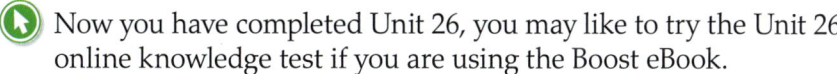 Now you have completed Unit 26, you may like to try the Unit 26 online knowledge test if you are using the Boost eBook.

27 Linear functions and solving simultaneous linear equations

- Use knowledge of coordinate pairs to construct tables of values and plot the graphs of linear functions, including where y is given implicitly in terms of x ($ax + by = c$), and quadratic functions of the form $y = x^2 \pm a$.
- Understand that straight-line graphs can be represented by equations. Find the equation in the form $y = mx + c$ or where y is given implicitly in terms of x (fractional, positive and negative gradients).
- Understand that the solution of simultaneous linear equations:
  - is the pair of values that satisfy both equations
  - can be found algebraically (eliminating one variable)
  - can be found graphically (point of intersection).

Plotting functions

Plotting graphs of functions usually involves identifying some of the points on the line first and then drawing either a straight line or a curve through the points. Knowing what type of function it is will also avoid the possibility of drawing the wrong type of line through the points.

You will already have seen in Unit 24 that linear functions can take more than one form. In general, these are the $y = mx + c$ form and the $ax + by = c$ form.

> **KEY INFORMATION**
>
> In the $ax + by = c$ form, it is usual for the constants a, b and c to be given as integer values.
>
> In the $y = mx + c$ form, the constants m and c can take fractional values.

They can be identified as linear functions (i.e. produce a straight line) because both x and y are raised to the power of 1, i.e. x^1 and y^1.

Worked examples

1 Plot the function $y = -\frac{1}{2}x + 2$.

Draw a table of results, by choosing some values of x.

x	y
−4	4
−2	3
0	2
2	1
4	0

Plot the points and draw a line through them.

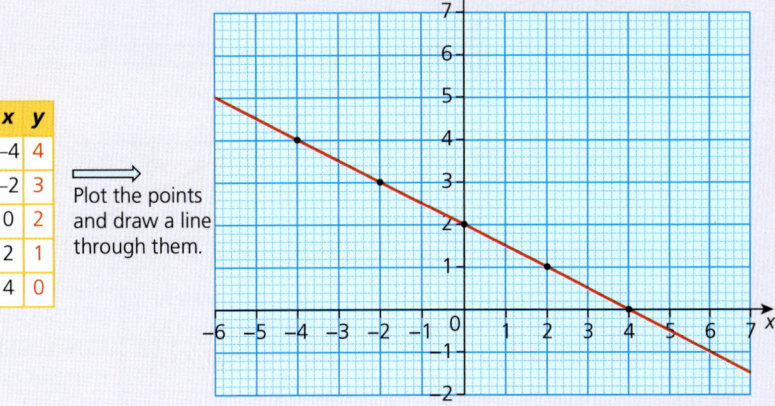

The straight line can be extended beyond the points at either end.

2 Plot the function $2y + x = 4$.

This can be plotted using the same method as above. However, each time it will involve rearranging the equation to find the value of y. A more efficient way, when the function is written in this format, is to find where it intersects both axes.

When any line intersects the y-axis, the x-coordinate is zero.

The value $x = 0$ is therefore substituted into the equation as follows:

$$2y + 0 = 4$$

Therefore $y = 2$, i.e. the line passes through the point (0, 2).

When any line intersects the x-axis, the y-coordinate is zero.

The value $y = 0$ is therefore substituted into the equation as follows:

$$2(0) + x = 4$$

Therefore $x = 4$, i.e. the line passes through the point (4, 0).

These can be plotted and a straight line drawn through them as shown.

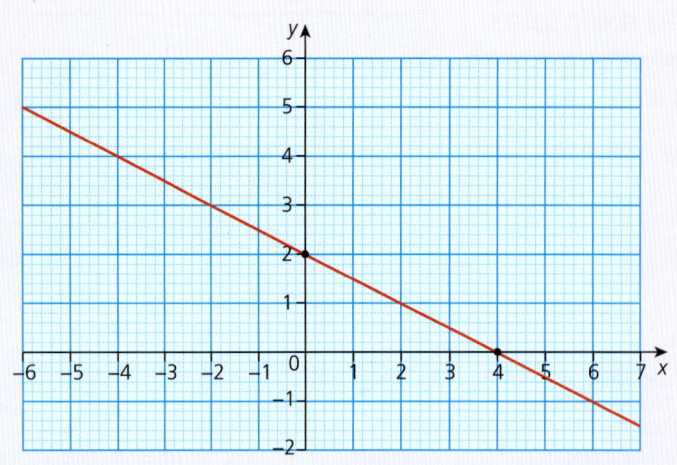

You may have noticed that both graphs are in fact the same. This is because $2y + x = 4$ is simply $y = -\dfrac{1}{2}x + 2$ rearranged into the $ax + by = c$ form.

Although only two points are needed to draw a straight line, it is good practice to check that a third point on the line fits the equation. It can be seen from the graph that (2, 1) lies on the line. Substituting this into the equation gives:

$$2(1) + 2 = 4$$
$$2 + 2 = 4$$
$$4 = 4$$

LET'S TALK

How could you prove algebraically that the point (1, 1) does not lie on the line?

As this is correct, the point (2, 1) lies on the line and the graph is correct.

3 Plot the function $y = x^2 - 3$.

By looking at the function, it can be deduced that, as the x is raised to a power of 2, this is not a linear function but is a quadratic one instead. A 'U' shaped curve will need to be drawn through the points.

A table of results can be drawn up as shown.

x	y
-2	1
-1	-2
0	-3
1	-2
2	1

Plot the points and draw a smooth curve through them.

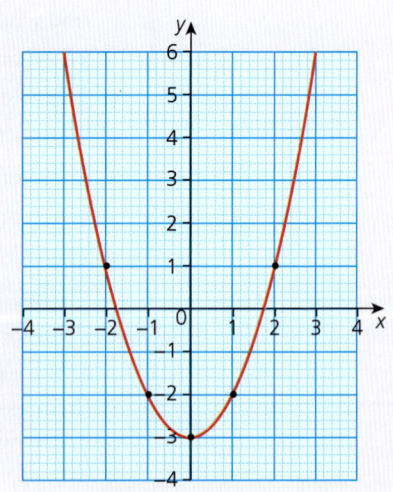

Exercise 27.1

 1 Six functions are given below.
Classify the functions into linear and quadratic functions.

$x - 2y = 4$ $\qquad$ $y = x^2 + 8$

$y = -\dfrac{1}{2}x + 2$ $\qquad$ $y - x^2 + 4 = 0$

$3y - \dfrac{1}{2}x + 2 = 0$ $\qquad$ $3y + 4x = 6$

2 In each of the following, the equation of a function is given.
i) Copy and complete the table of results for each function.
ii) Plot the graph of each function on a separate pair of axes.

a $y = \dfrac{1}{4}x + 6$

x	−8	−4	0	4	6	8
y						

b $y = -\dfrac{1}{3}x - 2$

x	−9	−6	−3	0	3	8
y			−1			

c $y = x^2 + 5$

x	−3	−2	−1	0	1	2	3
y		9					

d $5y + x = 5$

x	−10	−5	0			
y					−2	$-\dfrac{13}{5}$

 3 Two functions are given as $y = x^2 + 2$ and $2x + y = 5$.
a If the two functions were plotted, identify which of the coordinates below would belong to which line.

(−3, 11) $\quad$ (−2, 6) $\quad$ (0, 5) $\quad$ (−2, 9) $\quad$ (1, 3) $\quad$ (3, 11) $\quad$ (4, −3) $\quad$ (0, 2) $\quad$ (−4, 13)

b i) If the functions were graphed on the same axes, which of the points above would be at the intersection of the two lines?
ii) Justify your answer to part (i).

4 **a** Match each of the graphs A–D below to one of the following four functions.

$$4x + 3y = 12 \qquad 5x - 2y = -2 \qquad 3x + 4y = -2 \qquad 4x + 7y = 7$$

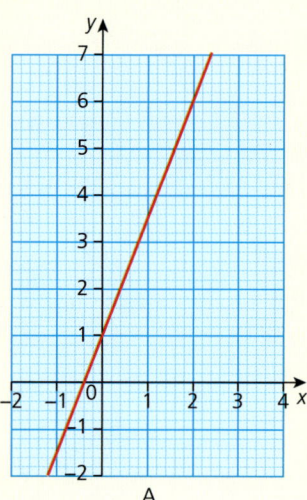

A

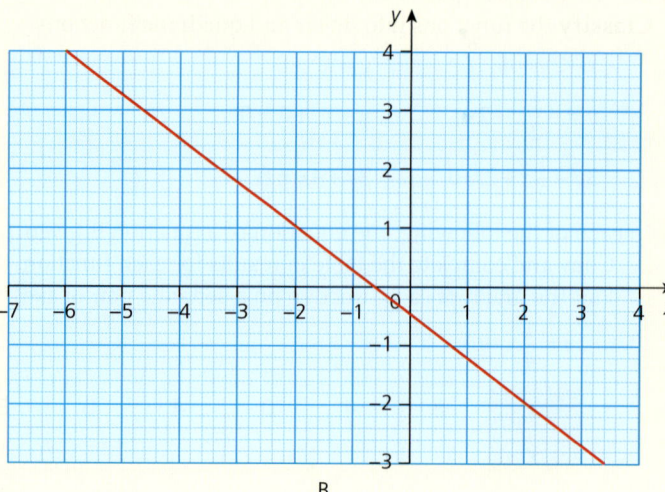

B

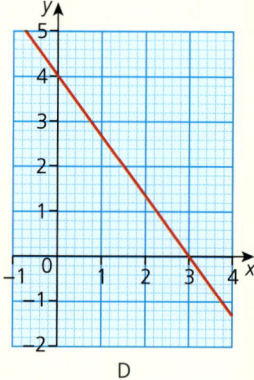

C

D

b i) Prove that the line $3x + 4y = -2$ crosses the y-axis at $\left(0, -\dfrac{1}{2}\right)$.

ii) Prove that it crosses the x-axis at $\left(-\dfrac{2}{3}, 0\right)$.

c Deduce where the line $4x + 7y = 7$ intersects each axis.

If m is positive, the straight line will slope this way:

If m is negative, the straight line will slope this way:

Deducing equations from graphs

So far you have studied how to plot a graph from the equation of a function.

It is of course possible to do the opposite, that is, deduce the equation of a straight line from its graph.

Recap

In Stage 8 you saw that every straight line takes the form $y = mx + c$, where m represents the gradient and c represents the intercept with the y-axis.

Comparing, say, $y = 2x + 1$ and $y = 4x + 1$ we can deduce a number of properties of the two lines.

Comparing with $y = mx + c$ it can be seen that they must both intersect the y-axis at the same point, i.e. (0, 1), also that $y = 4x + 1$ is steeper than $y = 2x + 1$ as its m value is greater.

This is confirmed in the graph on the right.

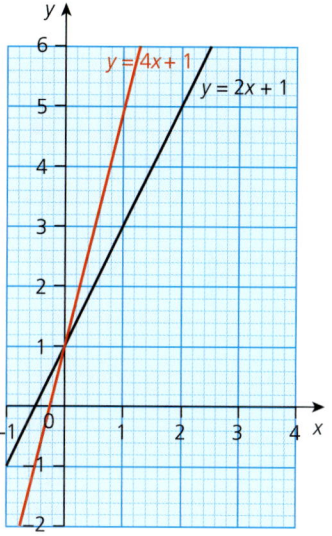

Similarly, comparing $y = 3x + 2$ and $y = 3x - 4$ we can deduce that as their gradients are the same, then they must be parallel lines, but they intersect the y-axis at different points.

$y = 3x + 2$ intersects it at (0, 2) and $y = 3x - 4$ at (0, −4), as shown below.

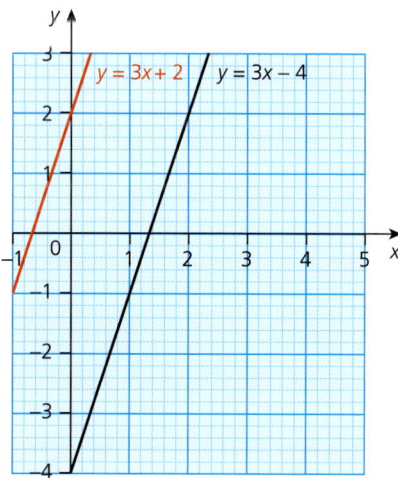

Calculating the gradient

The gradient of a straight line can be calculated by selecting any two points on the line and working out the vertical height between them and the horizontal distance between them.

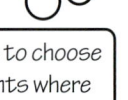

It is best to choose two points where the coordinates are easily identified.

For example, on the line $y = 3x - 4$ on the previous page, two clear points can be identified as (2, 2) and (0, −4).

Label one of them as point 1 and the other as point 2 as summarised in the diagram here.

LET'S TALK

Will the calculated value of the gradient be affected if the labelling of the points 1 and 2 is swapped?

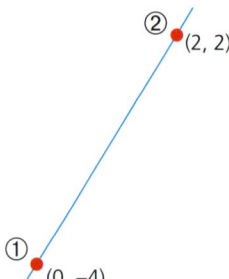

The gradient $= \dfrac{y_2 - y_1}{x_2 - x_1}$

$\qquad\quad = \dfrac{2 - -4}{2 - 0}$

$\qquad\quad = \dfrac{6}{2} = 3$

The gradient of the line $y = 3x - 1$ is calculated as 3.

Therefore, when a line is given in the form $y = mx + c$, the value of m is not just an indicator of steepness, it is the actual numerical value of the gradient.

> ### Worked examples

1 Deduce the equation of the following line, giving the answer in the form $ax + by = c$.

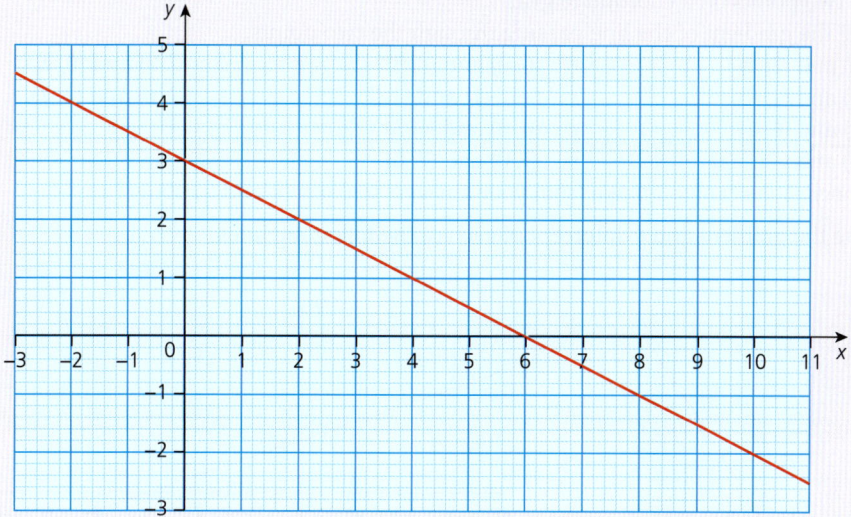

Choose two easily identifiable points, e.g. (0, 3) and (6, 0), and decide which is labelled ① and which is labelled ②. For example, (0, 3) is point ① and (0, 6) is point ②.

$$\text{Gradient} = \frac{y_2 - y_1}{x_2 - x_1} = \frac{0 - 3}{6 - 0} = -\frac{1}{2}$$

The y-intercept is shown as $+3$, therefore the equation of the line in the form $y = mx + c$ can be written as $y = -\frac{1}{2}x + 3$.

To rearrange in the form $ax + by = c$:

$$y = -\frac{1}{2}x + 3$$

$2y = -x + 6$ *Multiply both sides by 2.*

$x + 2y = 6$ *Add x to both sides.*

Therefore, the equation of the line is $x + 2y = 6$.

2 Calculate the gradient and y-intercept of the line with equation $3x + 4y = 8$.

It is easier to work out the gradient and y-intercept when the equation is in the form $y = mx + c$.

Rearranging $3x + 4y = 8$:

$4y = -3x + 8$ *Subtract 3x from both sides.*

$y = -\frac{3}{4}x + 2$ *Divide both sides by 4.*

Therefore, the gradient is $-\frac{3}{4}$ and the y-intercept is $+2$.

Exercise 27.2

1 For each of parts (a)–(j):
   i) plot the two points on a coordinate grid
   ii) calculate the gradient of the line passing through the two points.

a	(5, 4) and (2, 1)	e	(6, 0) and (0, 3)	i	(7, 2) and (0, 2)
b	(4, 6) and (1, 0)	f	(−1, −1) and (1, 7)	j	(2, 0) and (2, 7)
c	(0, 1) and (6, 4)	g	(0, 5) and (6, 3)		
d	(2, 4) and (−1, −2)	h	(4, 0) and (2, 6)		

   k What conclusions can you make from your answers to parts (i) and (j) above?

2 a Calculate the gradient and y-intercept of each of the following lines.

   i)   **ii)**

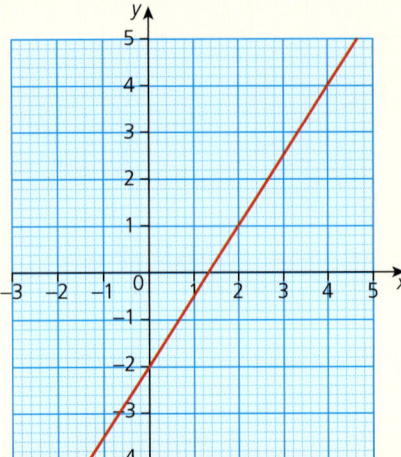   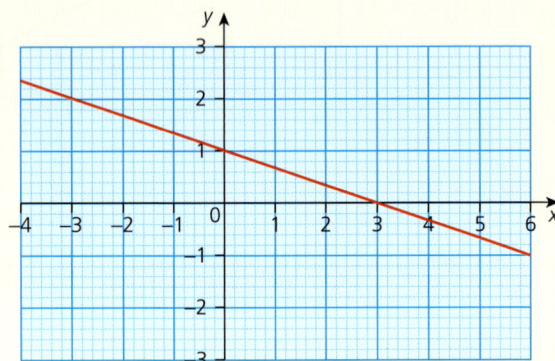

   b Use your answers to part (a) to write down the equation of each line in the form $y = mx + c$.

For the straight line represented by each equation in questions 3–5, find the value of:

   i) the gradient   **ii)** the y-intercept.

3 a $y = 2x + 1$ **d** $y = x$ **g** $y = -x + 4$

   b $y = 3x - 1$ **e** $y = x - \dfrac{1}{2}$ **h** $y = -x$

   c $y = \dfrac{1}{2}x - 3$ **f** $y = -3x + 4$

4 a $y - 2x = 4$ **d** $y + 2x = 4$ **g** $y - 5x + 4 = 0$

   b $y - x = -2$ **e** $y + 3x = -1$ **h** $y + 2x - 1 = 3$

   c $y - 3x = 0$ **f** $y - 1 - x = 0$

5 a $2y = 2x + 4$ **d** $5y = 5x$ **g** $\dfrac{1}{4}y = 1$

   b $2y = 4x - 2$ **e** $\dfrac{1}{2}y = 2x - 4$ **h** $2y - x + 6 = 0$

   c $3y = 9x + 3$ **f** $\dfrac{1}{3}y = x - 1$

6 The four tables below contain information about four different straight lines.
Match the correct properties for each line.

Equation in the form $y = mx + c$	Equation in the form $ax + by = c$	Gradient	y-intercept
$y = -\dfrac{2}{3}x + 1$	$6x - 2y = 1$	$\dfrac{2}{3}$	$+2$
$y = \dfrac{1}{3}x + 2$	$2x - 3y = 3$	$\dfrac{1}{3}$	$-\dfrac{1}{2}$
$y = \dfrac{2}{3}x - 1$	$2x + 3y = 3$	$-\dfrac{2}{3}$	-1
$y = 3x - \dfrac{1}{2}$	$x - 3y = -6$	3	1

 7 a A straight line is shown.

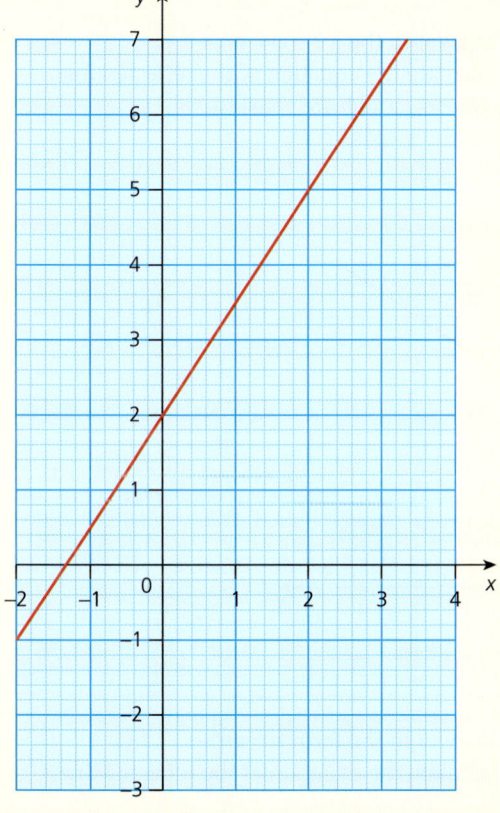

i) Calculate its equation, giving your answer in the form $y = mx + c$.
ii) The line is reflected in the y-axis. Calculate the equation of the reflected line.

b   Another straight line is shown.

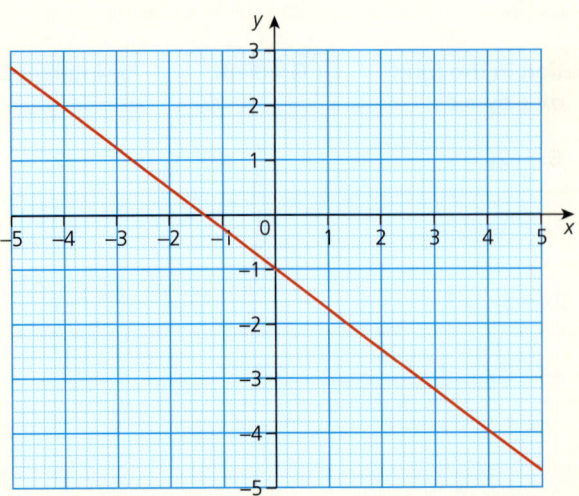

i)   Calculate its equation, giving your answer in the form $y = mx + c$.
ii) The line is reflected in the y-axis. Calculate the equation of the reflected line.

c   A straight line has equation $y = ax + b$, where a and b are constants.
i)   Deduce its equation when reflected in the y-axis.
ii) Justify your answer to part (i).

8 a   A straight line is shown.

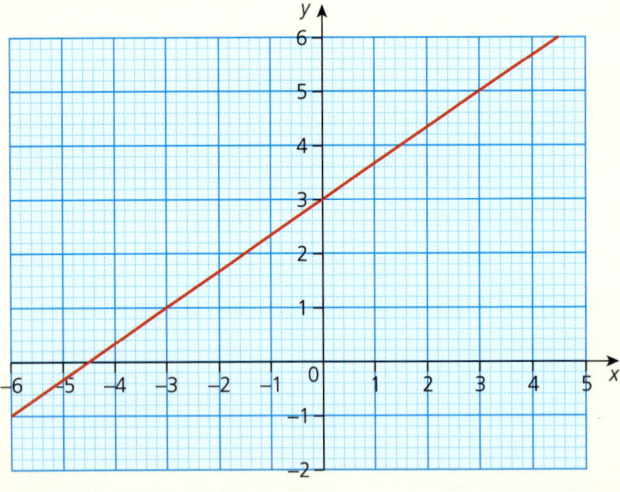

i)   Calculate its equation, giving your answer in the form $y = mx + c$.
ii) The line is reflected in the x-axis. Calculate the equation of the reflected line.

b Another straight line is shown.

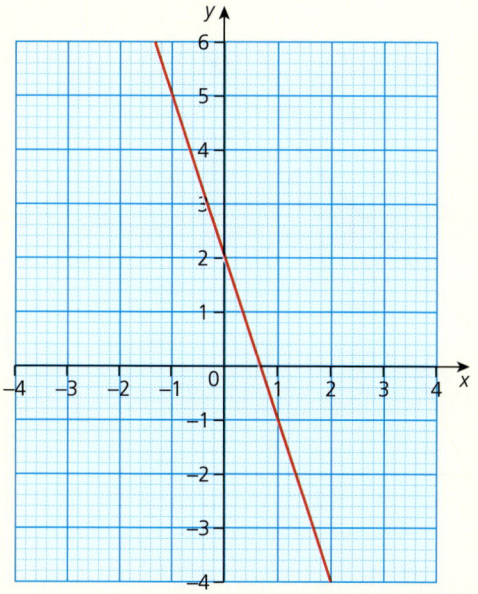

i) Calculate its equation, giving your answer in the form $y = mx + c$.
ii) The line is reflected in the x-axis. Calculate the equation of the reflected line.
c A straight line has equation $y = ax + b$, where a and b are constants.
i) Deduce its equation when reflected in the x-axis.
ii) Justify your answer to part (i).

Solving simultaneous equations graphically

The coordinates of all points on a straight line will fit the equation of that straight line.

For example, for the line with equation $y = x$, the coordinates of all the points on that line must satisfy that equation. The points (1, 1), (2, 2), (5, 5), (−2, −2) etc. will all lie on the line because in each case the y-coordinate is equal to the x-coordinate, i.e. $y = x$.

Similarly, the points (1, −3), (2, −1), (5, 5), (−2, −9) will all lie on the line with equation $y = 2x − 5$.

You can see from the coordinates written above, that the point with coordinates (5, 5) lies on both lines.

If both lines are plotted on the same axes, the following graph is produced.

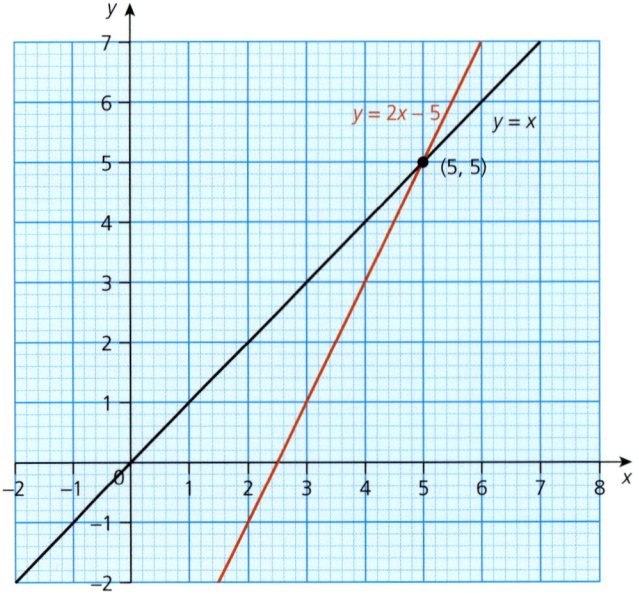

The point (5, 5), as it lies on both lines, occurs at their point of intersection. Working with two equations and finding a solution that satisfies both is known as solving **simultaneous equations**.

Worked example

Plot the following lines and find their point of intersection:

$$x - 2y = 8 \quad \text{and} \quad y = -x + 2$$

The plots and point of intersection are as shown.

The answer of (4, −2) can be checked by substituting it into both equations.

In $x - 2y = 8$:

$4 - 2(-2) = 8$, which leads to $8 = 8$. ✔

In $y = -x + 2$:

$-2 = -(4) + 2$, which leads to $-2 = -2$. ✔

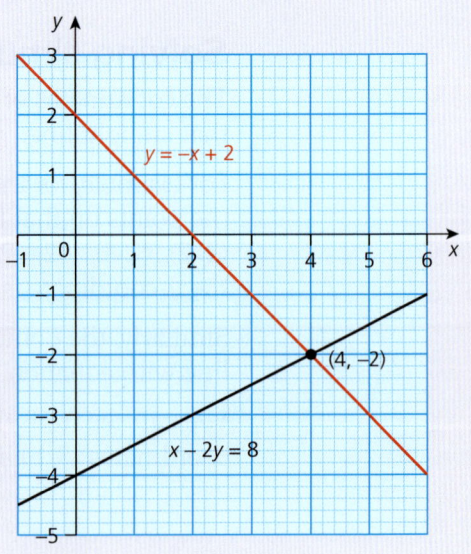

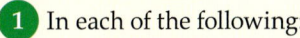

Exercise 27.3

1 In each of the following:

i) plot both lines on the same axes and where possible find the coordinates of their point of intersection

ii) check your answer.

    a $y = x + 2$ and $y = -\frac{1}{2}x + 5$

    b $x - 3y = 2$ and $x + y = 2$

    c $y = -\frac{1}{2}x + 4$ and $x + 2y = 3$

 2 **a** Two straight lines intersect at the point with coordinates (3, 6). Give possible equations for the two straight lines.

    b Show that your answer to part (a) is correct by plotting both lines.

    c Give another pair of possible equations for two lines that intersect at (3, 6).

3 The equations of three lines are given as $y = -2x + 5$, $4y + x = -8$ and $y = \frac{3}{2}x - 2$.

Find the area of the triangle enclosed by the three lines.

> **LET'S TALK**
>
> How many pairs of lines can intersect at (3, 6)?

Approximate solutions

Plotting straight lines and seeing where they intersect is a good method if the coordinates of the point of intersection are whole numbers. If they are decimals, then it may be difficult to read them off accurately from the graph. If this is the case, then the solutions are only approximate.

Worked example

Solve these simultaneous equations by drawing a graph:

$$y = \frac{1}{2}x - 2 \quad \text{and} \quad y = -2x + 2$$

First draw the straight line $y = \frac{1}{2}x - 2$ on a coordinate grid.

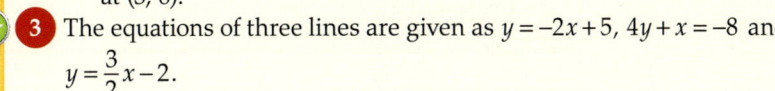

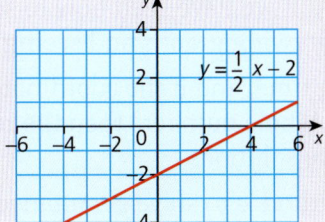

> The x and y values of all the points along this line satisfy the equation $y = \frac{1}{2}x - 2$

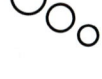

The x and y values of all the points along the blue line satisfy the equation $y = -2x + 2$

Now draw the line $y = -2x + 2$ on the same axes.

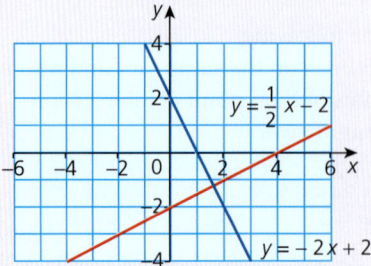

At the point where the two graphs intersect, both equations are satisfied.

From the graph we can see that the coordinates of this point are approximately (1.6, −1.2).

So the approximate solution of the pair of simultaneous equations is $x = 1.6$, $y = -1.2$.

Exercise 27.4

For each of these pairs of equations:
a   draw the two straight lines on the same coordinate grid
b   use the coordinates of the point of intersection to find an approximate solution to the simultaneous equations.

1 $y = x + 3$ and $y = -2x + 1$

2 $y = \dfrac{1}{3}x - 1$ and $y = -x + 1$

3 $y = 2x - 3$ and $y = \dfrac{1}{2}x + 1$

4 $y + 2x - 3 = 0$ and $2y = x - 6$

5 $3y + x = 3$ and $y + 3x + 3 = 0$

Solving simultaneous equations algebraically

In the previous section we have seen how to solve two equations simultaneously by drawing graphs and seeing where they intersect.

This can take a long time, and if the coordinates of the point of intersection are not whole numbers then the results from the graph are only approximate solutions.

A more accurate method is to solve simultaneous equations algebraically.

LET'S TALK

Is there a pair of values that works in both equations?

For example, solve the following pair of simultaneous equations:

$3p + q = 17$ equation ①

$5p - q = 15$ equation ②

How can we find a solution which satisfies both equations?

Try $p = 5$, $q = 2$ in equation ①:

$3p + q = 17$

$3 \times 5 + 2 = 17$ ✔

Try $p = 5$, $q = 2$ in equation ②:

$5 \times 5 - 2 = 15$

$25 - 2 = 15$

$23 = 15$ ✘

The value of the left-hand side does not equal the right-hand side, so $p = 5$ and $q = 2$ is not a solution that works for both equations.

There is only one solution that satisfies both equations simultaneously. The method of trying lots of different possible solutions is called **trial and improvement**. It works but it can take a long time, especially if the solutions are not whole numbers.

A better method is called the method of elimination.

Worked examples

1   Solve the two simultaneous equations below.

$3p + q = 17$ equation ①

$5p - q = 15$ equation ②

If the two equations are added, one of the variables (q) is eliminated:

$3p + q = 17$

$\underline{5p - q = 15}$

$8p = 32$

Therefore $p = 4$.

To find the value of q, substitute $p = 4$ into equation ①:

$3p + q = 17$

$(3 \times 4) + q = 17$

$12 + q = 17$

$q = 17 - 12$

$q = 5$

So $p = 4$ and $q = 5$.

KEY INFORMATION

$p = 4$ can be substituted into either of the two equations in order to find the value of q.

It is important to substitute both of your solutions into the other equation to check that they are correct.

Check by substituting $p = 4$ and $q = 5$ in equation ②:

$$5p - q = 15$$
$$20 - 5 = 15 \quad ✔$$

In that last example one variable was eliminated by addition. However, there are times when a variable can be eliminated by subtraction.

2　Solve these simultaneous equations by eliminating by subtraction.

$$5a + 2b = 16 \quad \text{equation ①}$$
$$3a + 2b = 12 \quad \text{equation ②}$$

Subtract equation ② from equation ①:

$$5a + 2b = 16$$
$$\underline{3a + 2b = 12}$$
$$2a = 4$$

Therefore $a = 2$.

Substitute $a = 2$ in equation ①:

$$5a + 2b = 16$$
$$10 + 2b = 16$$
$$2b = 16 - 10$$
$$2b = 6$$
$$b = 3$$

Check by substituting $a = 2$, $b = 3$ in equation ②:

$$3a + 2b = 12$$
$$6 + 6 = 12 \quad ✔$$

3　Solve these simultaneous equations by eliminating by subtraction.

$$4c - 3d = 14 \quad \text{equation ①}$$
$$c - 3d = -1 \quad \text{equation ②}$$

Subtract equation ② from equation ①:

$$4c - 3d = 14$$
$$\underline{c - 3d = -1}$$
$$3c = 15$$

Therefore $c = 5$.

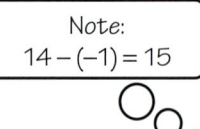

Note:
$14 - (-1) = 15$

Substitute $c = 5$ in equation ①:

$$4c - 3d = 14$$
$$20 - 3d = 14$$
$$-3d = 14 - 20$$
$$-3d = -6$$
$$d = 2$$

Check by substituting $c = 5$, $d = 2$ in equation ②:

$$c - 3d = -1$$
$$5 - 6 = -1 \quad ✔$$

LET'S TALK

When is elimination by addition used and when is subtraction used?

Exercise 27.5

Solve each of these pairs of simultaneous equations by elimination.

1 $a + b = 8$
$3a - b = 12$

2 $c + d = 7$
$4c - d = 8$

3 $5e - 2f = 4$
$2e - 2f = -2$

4 $4g - 2h = 8$
$g - 2h = -1$

5 $j + k = 7$
$4j - k = 23$

6 $l - 2m = -5$
$3l - 2m = -3$

7 $2u + v = 6$
$3u + v = 7$

8 $3a + 2b = 10$
$3a - b = 4$

9 $c + 4d = 16$
$5c - 4d = 8$

10 $e + 2f = 8$
$5e - 2f = 4$

11 $g + h = 3$
$g + 3h = 7$

12 $j + 2k = 8$
$j - 3k = -7$

13 $3l + 2m = 14$
$5l - 2m = 2$

14 $n + 4p = 13$
$12 - 4p = 0$

15 $3r - s = -1$
$3r - 2s = -8$

16 $t + u = 11$
$t + 3u = 29$

Harder cases

Consider these two simultaneous equations.

$x + 2y = 7$ equation ①

$3x + y = 11$ equation ②

In this case, simply adding or subtracting the equations will not result in the elimination of either x or y. An extra process is needed.

If all the terms of an equation are multiplied by the same number, it will still balance. For example, multiplying $x + 2y = 7$ throughout by 3 gives the equation $3x + 6y = 21$.

This is equivalent to $x + 2y = 7$.

Worked example

Solve these two equations simultaneously.

$x + 2y = 7$ equation ①

$3x + y = 11$ equation ②

Multiply equation ① throughout by 3 and call the result equation ③:

$x + 2y = 7$ equation ①

$\times 3$ $3x + 6y = 21$ equation ③

Now replace equation ① in the pair of simultaneous equations by equation ③:

$$3x + 6y = 21 \qquad \text{equation ③}$$
$$3x + y = 11 \qquad \text{equation ②}$$

Subtract equation ② from equation ③ to eliminate x:

$$3x + 6y = 21$$
$$\underline{3x + y = 11}$$
$$5y = 10$$

Therefore $y = 2$.

Substitute $y = 2$ in equation ①:

$$x + 2y = 7$$
$$x + 4 = 7$$
$$x = 7 - 4$$
$$x = 3$$

Check by substituting $x = 3$, $y = 2$ in equation ②:

$$3x + y = 11$$
$$9 + 2 = 11 \quad ✔$$

Exercise 27.6

Solve each of these pairs of simultaneous equations by elimination.

1. $a + 3b = 5$
   $3a + 2b = 8$

2. $b + 3c = 7$
   $2b + c = 4$

3. $3m - 4n = -14$
   $2m + 2n = 14$

4. $p + q = 7$
   $2p - 3q = -6$

5. $r + 2s = 0$
   $3r - s = -7$

6. $2t + u = 1$
   $3t - 2u = 0$

7. Calculate the lengths of the sides of the rectangle shown.

You will need to use your **conjecturing** skills to solve the rest of the questions in this exercise.

8 Calculate the lengths of the sides of the rectangle shown.

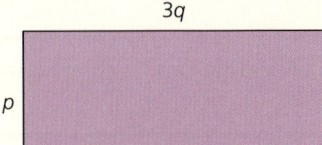

3q

p   *5 − q*

13 − 2p

LET'S TALK

With a friend, can you make more of these types of puzzles and get each other to work them out?

 9 For each of the following problems, form two equations and solve them to find the two numbers.

a The sum of two numbers is 50, and their difference is 18.

b The sum of two numbers is 60, and half of their difference is 10.

10 Two families enter a café and buy cups of coffee and cups of tea. The first family buys four coffees and six teas and the bill totals $14.60. The second family buys two coffees and seven teas and the bill totals $12.50.

Calculate the price of:

a one coffee

b twelve coffees and eight teas.

11 Clay tiles are used to form a path. Each tile has dimensions $m \times n$, where $m > n$.

The tiles are laid in blocks of 16 tiles. One block and two blocks of tiles are shown below.

 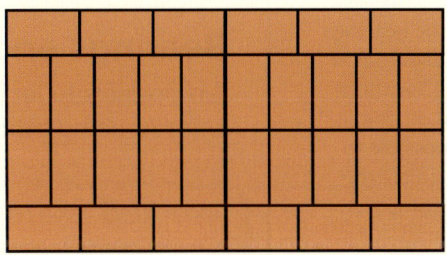

LET'S TALK

Can the values of *m* and *n* be calculated using ratios? If so, how?

The perimeter of one block of tiles is 186 cm, whilst the perimeter of two blocks is 276 cm.

a Form two equations in terms of *m* and *n* from the information given.

b Calculate the lengths *m* and *n*.

 Now you have completed Unit 27, you may like to try the Unit 27 online knowledge test if you are using the Boost eBook.

- Use knowledge of bearings and scaling to interpret position on maps and plans.

Bearings

Recap

In Stage 8 you were introduced to the topic of bearings. A bearing is the direction or position of something in relation to a fixed point. The direction is given as an angle, measured clockwise from North and written as a three-digit number.

For example, two points A and B are shown here. To measure the bearing of B from A, the following steps are taken:

- Draw a straight line joining A and B.
- Draw a North arrow at A.
- In a clockwise direction measure the angle from North to the line AB.
- Write the angle as a three-digit number.

Therefore, the bearing of B from A is 150°.

Bearings and distances

It is possible to locate the position of a second point B from a fixed point A, using bearings and distances.

> ### Worked example
>
> A point B is 3 cm from A on a bearing of 225°.
>
> Draw a scale drawing to show the positions of A and B.
>
> This can be done by following these steps:
> - Mark a point A and draw a North arrow from it.
>
>

LET'S TALK

If the bearing of B from A is 150°, without measuring, is it possible to calculate the bearing of A from B?

- Draw a line from A at an angle of 225° measured clockwise from North. Every point on the line is on a bearing of 225° from A.

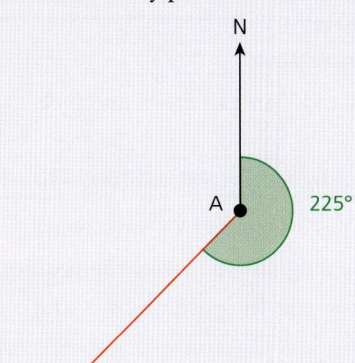

- Measure 3 cm from A along the line and mark the point B.

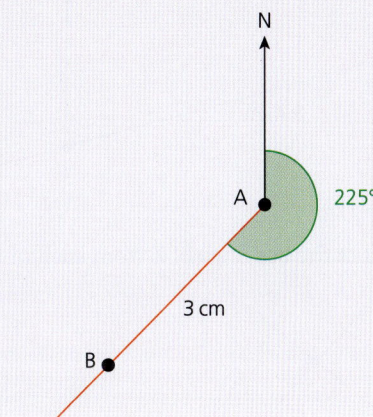

Similarly it is possible to locate the position of a point using its bearing from two other fixed points.

Worked example

Two lifeboats X and Y are out at sea and pick up a distress signal from a ship Z.

The bearing of Z from X is 052°, whilst the bearing of Z from Y is 317°.

The distance XY is 8.4 km and Y is on a bearing of 105° from X.

a   Draw a scale diagram and locate the position of Z.

First draw, to scale, the positions of X and Y relative to each other.

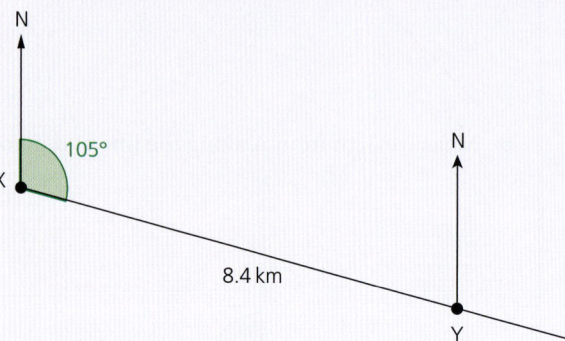

Scale 1 cm = 1 km

Then construct the bearings of Z from each of the points X and Y.

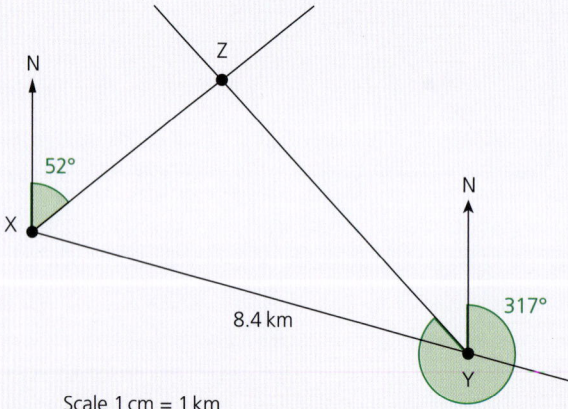

Scale 1 cm = 1 km

b   The closest lifeboat goes to rescue Z. From your scale drawing, work out the distance it needs to travel to reach boat Z.

The closest lifeboat to Z is X.

By measuring the diagram it can be worked out that the distance XZ is 4.5 km (to 1 d.p.).

Exercise 28.1

For questions 1 and 2, draw diagrams using a scale of 1 cm : 1 km.
Take North as a line vertically up the page.

1 **a** A boat starts at a point A. It travels a distance of 7 km on a bearing of 135° to point B.
From B it travels 12 km on a bearing of 250° to point C.
Draw a diagram to show these bearings and journeys.

b The boat makes its way straight back from C to A.
What distance does it travel and on what bearing?

c Another boat travels directly from A to C.
What are the distance and bearing of this journey?

2 **a** An athlete starts at a point P. He runs on a bearing of 225° for a distance of 6.5 km to point Q. From Q he runs on a bearing of 105° a further distance of 7.8 km to a point R.
From R he runs towards a point S a further distance of 8.5 km and on a bearing of 090°.
Draw a diagram to show these bearings and journeys.

b Calculate the distance and bearing the athlete has to run to get directly from S back to P.

3 The map extract shows a part of Malaysia and Singapore.
The scale of the map is 1 : 4 000 000.

a A tourist travels from Kuala Lumpur to Singapore. By measuring the map with a ruler, calculate the real (direct) distance between the two cities. Give your answer in kilometres.

b What is the bearing from Kuala Lumpur to Singapore?

c   A traveller decides to travel from Kuala Lumpur to Singapore and then on to Kuantan, before returning to Kuala Lumpur. Copy and complete this table of distances and bearings.

Journey	Distance (km)	Bearing
Kuala Lumpur to Singapore		
Singapore to Kuantan		
Kuantan to Kuala Lumpur		

4 A light aeroplane flies from London to Cambridge and then on to Birmingham and Cardiff before returning to London again. The cities are shown on this 1:4 000 000 scale map of part of Britain.

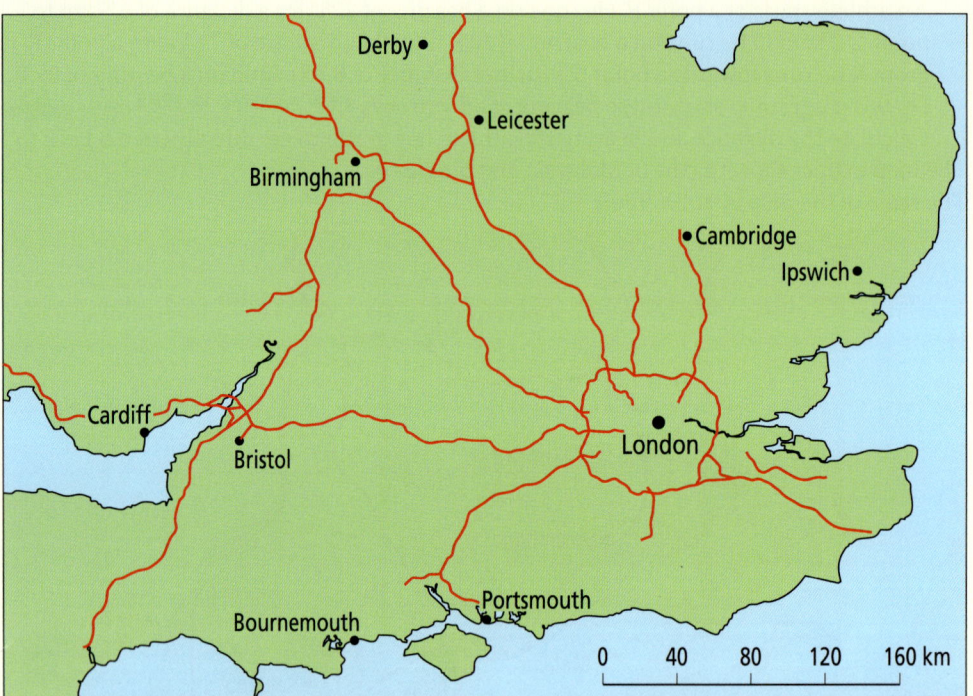

Copy and complete the table below by calculating the true distance between the cities and the bearing for each stage of the journey.

Journey	Distance (km)	Bearing
London to Cambridge		
Cambridge to Birmingham		
Birmingham to Cardiff		
Cardiff to London		

⑤ This 1 : 10 000 000 scale map shows part of southern Africa.

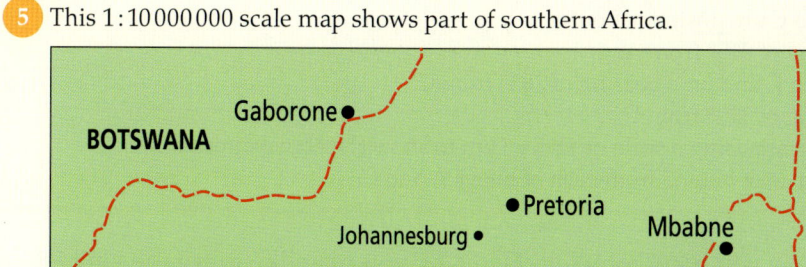

A businesswoman needs to visit the cities of Pretoria, Mbabne, Maputo, Maseru and Gaborone.
The only conditions of the trip are that she must start in Pretoria and end in Maseru.
Copy the table below and plan a possible route for the businesswoman by filling in the details of the route that she takes.

Journey	Distance (km)	Bearing
Pretoria to _____		
_____ to _____		
_____ to _____		
_____ to _____		
_____ to Maseru		

⑥ Two lionesses L_1 and L_2 are hunting down a gazelle G. L_1 and L_2 are 120 m apart and L_2 is on a bearing of 072° from L_1. The bearing of the gazelle from L_1 is 155°, whilst the bearing of the gazelle from L_2 is 208°.

a   Draw a scale diagram to show the above information.

b   From your diagram deduce the distance of the gazelle from each of the lionesses.

Make sure the scale used is clearly shown.

7 Three buoys A, B and C are positioned out at sea.

Buoy B is on a bearing of 015° from A.

Buoy A is 6.6 km from C and on a bearing of 283° from C.

Buoy B is on a bearing of 312° from C.

   a Draw a scale diagram using a scale of 1 cm = 1 km to show the above information.

   b Deduce the distance of buoy B from each of buoys A and C.

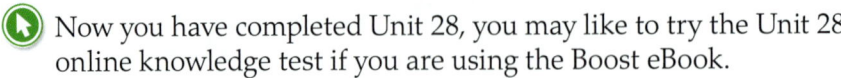 Now you have completed Unit 28, you may like to try the Unit 28 online knowledge test if you are using the Boost eBook.

29 Direct and inverse proportion

- Understand the relationship between two quantities when they are in direct or inverse proportion.
- Use knowledge of ratios and equivalence for a range of contexts.

Ratio

You already know from Stages 7 and 8 that a ratio shows the relative sizes of two numbers. It can be expressed in different ways.

Recap

> **KEY INFORMATION**
>
> Simplifying a ratio is done in the same way as simplifying a fraction. That is, divide both parts by the highest common factor.

Worked example

Zachary has a collection of toy cars.

He has 12 red cars, 9 silver cars and 4 blue cars.

a   What is the ratio of silver cars to red cars?

As a ratio	silver : red	$= 9:12$	$= 3:4$
As a fraction	$\dfrac{\text{silver}}{\text{red}}$	$= \dfrac{9}{12}$	$= \dfrac{3}{4}$
As a percentage	$\dfrac{\text{silver}}{\text{red}} \times 100$	$= \dfrac{9}{12} \times 100$	$= 75\%$

b   What is the ratio of red cars to the total number of cars?

The total number of cars is $12 + 9 + 4 = 25$.

As a ratio	red : total	$= 12:25$	
As a fraction	$\dfrac{\text{red}}{\text{total}}$	$= \dfrac{12}{25}$	
As a percentage	$\dfrac{\text{red}}{\text{total}} \times 100$	$= \dfrac{12}{25} \times 100$	$= 48\%$

To compare two different ratios, it is often helpful to write each ratio as a percentage.

> **Worked example**
>
> a   In a five-set tennis match, tennis player A hits 120 first serves. 84 of these first serves go in.
>
>   What is the success rate of player A's first serves?
>
> | As a ratio | $84:120$ | $= 7:10$ |
> |---|---|---|
> | As a fraction | $\dfrac{84}{120}$ | $= \dfrac{7}{10}$ |
> | As a percentage | $\dfrac{84}{120} \times 100$ | $= 70\%$ |
>
> b   In the same tennis match, player B hits 125 first serves and 90 go in.
>
>   What is the success rate of player B's first serves?
>
> | As a ratio | $90:125$ | $= 18:25$ |
> |---|---|---|
> | As a fraction | $\dfrac{90}{125}$ | $= \dfrac{18}{25}$ |
> | As a percentage | $\dfrac{90}{125} \times 100$ | $= 72\%$ |
>
> It is easy to see that player B has a higher success rate by looking at the percentages. It is not so obvious with the ratios or fractions.

To divide a quantity in a given ratio, fractions are useful.

> **Worked example**
>
> A piece of wood is 150 cm long. It is cut into two pieces in the ratio $2:3$.
>
> How long is each piece?
>
> $2:3$ gives 5 parts altogether.
>
> One piece of wood is made of 2 parts, which is $\dfrac{2}{5}$ of the total.
>
> $\dfrac{2}{5} \times 150\,\text{cm} = 60\,\text{cm}$
>
> The other piece is made of 3 parts, which is $\dfrac{3}{5}$ of the total.
>
> $\dfrac{3}{5} \times 150\,\text{cm} = 90\,\text{cm}$

Exercise 29.1

1. In basketball, player A scores 142 times out of 200 attempts from the free throw line.
   Player B scores 81 times out of 150 attempts and player C scores 76 times out of 120 attempts.
   a   What is each player's success rate:
      i)   as a ratio
      ii) as a percentage?
   b   Which player has the best success rate?

2. $4000 is divided between three children in the ratio $2:3:5$.
   How much does each child receive?

3. Peter is 150 cm tall. His friend Mehmet is 175 cm tall.
   What is the ratio of their heights in its simplest form?

4. Asli has a number of coins in a box. There are thirty-four 25 cent coins, eighty-five 10 cent coins, seventy-two 5 cent coins and nine 1 cent coins.
   As a percentage, write down the ratio of:
   a   25 cent coins to the total
   b   10 cent coins to 25 cent coins
   c   1 cent coins to 5 cent coins
   d   the total number of coins to 1 cent coins.

5. Gabriella and Gemma take penalties during a football practice.
   Gabriella's ratio of scoring to missing is $5:1$.
   Gemma's ratio of scoring to missing is $4:1$.
   Gabriella states that, as 5 is bigger than 4, she must have scored more penalties than Gemma.
   Give a **convincing** explanation to show whether Gabriella's statement is always true.

6. The ratio $a:b = 5:3$.
   If $a + b = 72$, calculate the value of $a - b$.

7. Three friends, Anna, Baniti and Chinara, are playing a board game. The board game involves players winning or losing cards. The total number of cards is fixed throughout the game.
   At the start of the game they each have the same number of cards.
   Midway through the game, Anna, Baniti and Chinara have the cards in the ratio $2:3:5$.
   At the end of the game Anna, Baniti and Chinara have them in the ratio $9:5:6$.
   a   Who had more cards at the end than they did at the start? Justify your answer.
   b   Did Baniti have more cards midway through the game or at the end? Justify your answer.
   c   Can Chinara have had 45 cards midway through the game? Justify your answer.

8. The side lengths of a triangle are as shown.
   The ratio of the lengths $XZ:YZ$ is $2:3$.
   a   Calculate the ratio of $XY:XZ$.
   b   i)   Is XYZ a right-angled triangle?
      ii) Justify your answer to part (i).

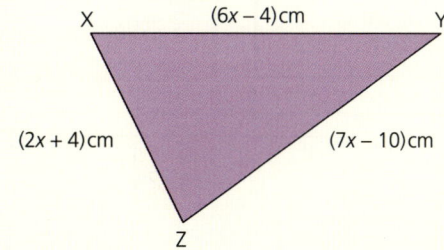

More complex problems involving ratios, require an understanding that ratios are a proportion rather than actual quantities. This, along with an ability to work 'backwards' through a problem, can be quite tricky.

Worked examples

1   Some sweets are shared in the ratio $3:7:5$ between X, Y and Z.

If Y gets 56 sweets, what is the total number of sweets (T)?

The total number has been divided into $3 + 7 + 5$ parts, i.e. 15 parts, of which Y receives 7 parts. Therefore, Y receives $\frac{7}{15}$ of the total, T.

This can be formed into the equation $\frac{7}{15} \times T = 56$.

Rearranging to find T:

$\quad 7T = 56 \times 15 \qquad$ *Multiply both sides by 15.*

$\quad 7T = 840$

$\quad\ \ T = 120 \qquad\qquad$ *Divide both sides by 7.*

Therefore, there were 120 sweets in total.

2   Two friends P and Q share prize money from a competition in the ratio $3:8$ respectively. If P receives \$30 less than Q, calculate the total amount of prize money that they shared.

Let the total be X.

The ratio implies that P receives $\frac{3}{11}$ of X and Q receives $\frac{8}{11}$ of X.

Therefore, as the difference between the two amounts is \$30, this can be written as an equation as:

$$\frac{8}{11}X - \frac{3}{11}X = 30$$

Therefore $\frac{5}{11}X = 30$

$\qquad\quad 5X = 330 \quad$ *Multiply both sides by 11.*

$\qquad\qquad X = 66 \quad$ *Divide both sides by 5.*

Therefore, the total prize money was \$66.

LET'S TALK

Can you think of another way to solve this problem?

LET'S TALK
In each of the questions in this exercise, there is more than one way of solving it. Discuss some different methods.

Exercise 29.2

 1 The ratio of boys to girls on a school bus at the start is $3:4$ respectively.

At the first stop, some more children get on the bus and the ratio of boys to girls is now $5:4$.

a For each of the following statements, decide whether it is:
- definitely true
- possibly true
- definitely false.

i) Only boys got on the bus at the first stop.

ii) There are more boys than girls on the bus after the first stop.

iii) There are 28 students on the bus at the start.

iv) There are 35 students on the bus after the first stop.

b Give a justification for your choice in part (a).

 2 An unopened sweet packet contains red and yellow sweets in the ratio $1:1$.

A child opens the packet and eats six of the red sweets.

The ratio of red to yellow sweets in the packet is now $4:5$.

How many sweets are still in the packet?

3 At a wedding, the ratio of men : women : children is $13:10:3$.

If there are 60 more men than children, calculate the number of women at the wedding.

4 Two triangles ABC and DEF are as shown.

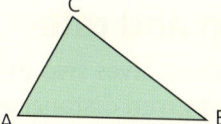 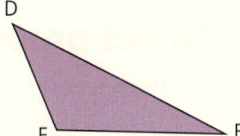

The ratio of the angles A : B : C is $5:4:9$, whilst the ratio of the angles D : E : F is $6:16:5$.

a Give a reasoned explanation why each triangle has an angle that is the same size as an angle in the other triangle.

b i) Which angle in ABC is the same size as an angle in DEF?

ii) Calculate the size of that angle.

5 A quadrilateral ABCD has angles in the ratio $2:3:8:5$.

Give a reasoned explanation why none of the angles differ from another angle by 50°.

6 Two numbers are in the ratio $3:10$.

If one of the numbers is 0.6, find two possible values for the other number.

7 The ratio $a:b = 3:4$, whilst the ratio $b:c = 7:2$.

Calculate the ratio of $a:c$ in its simplest form.

Direct proportion

If two ratios are equivalent they are in **direct proportion**.

> **Worked example**
>
> A picture frame measures 12 cm by 20 cm. A second frame measures 36 cm by 60 cm.
>
> Are the picture frames in proportion?
>
> The ratios of the dimensions of the two frames are 12 : 20 and 36 : 60.
>
> The ratios can be re-written as the fractions $\frac{12}{20}$ and $\frac{36}{60}$.
>
> $$\frac{12}{20} = \frac{3}{10} \quad \text{and} \quad \frac{36}{60} = \frac{3}{10}$$
>
> Alternatively, to test whether fractions are equivalent we can start with the assumption that they are equal, i.e. $\frac{12}{20} = \frac{36}{60}$.
>
> Cross-multiplying, which is equivalent to multiplying both sides by 60 and by 20, gives:
>
> $$12 \times 60 = 36 \times 20$$
> $$720 = 720$$
>
> Therefore, the ratios are equivalent and the picture frames are in the same proportion.

Direct proportion and rate

A rate is a proportion that involves two different units. For example, the time it takes to walk a given distance is a proportion involving time and distance, and the number of litres of petrol used to travel a given distance is a proportion involving volume and distance. They are examples of rates.

Note that the time of 1 hour 30 minutes has been changed to 1.5 hours. It could also have been changed to 90 minutes.

> **Worked example**
>
> Maria walks 6 km in 1 hour 30 minutes.
>
> How long will it take her to walk 10 km at the same rate?
>
> Let x be the time Maria takes to walk 10 km.
>
> Distance : Time
>
> 6 : 1.5
>
> 10 : x

The ratios have to be in proportion because she walks at the same rate. So:

$$\frac{6}{1.5} = \frac{10}{x}$$

$6x = 10 \times 1.5$ *Multiply both sides by x and 1.5.*

$6x = 15$

$x = \dfrac{15}{6} = 2.5$

It will take her 2 hours 30 minutes.

Exercise 29.3

Use direct proportion to work out these problems.

1 A car uses 20 litres of petrol to travel 160 km.
How much petrol does the car use to travel 400 km?

2 A 3 kg bag of potatoes costs $1.50. What does a 7 kg bag cost?

3 a In July 2020, £1 (pound sterling) was worth $1.25.
     i) How many pounds was $1200 worth?
     ii) How many dollars was £1200 worth?

    b At the same time, £1 was worth €1.10 (euros).
     i) How many euros was £250 worth?
     ii) How many pounds was €240 worth?

    c Using the information above, work out:
     i) how many dollars you would get for €2500
     ii) how many euros you would get for $2500.

4 A shop sells a liquid detergent in two sizes.
Size X was originally sold for $4.20 but
is now 20% off, whilst size Y is being sold for $5.80.
Which size is better value for money?
Give a **convincing** reason for your answer.

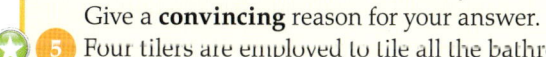
X 600 ml   Y 1000 ml

5 Four tilers are employed to tile all the bathrooms
of houses on a new housing estate.
It is estimated that there are 48 000 tiles to be tiled.
Each tiler can tile on average 220 tiles per day, and each receives
$120 per day.
   a Calculate the total that the four tilers will be paid for the job.
   b Calculate the total that the four tilers will be paid for the job using
     a different method of calculation to the one you used in part (a).

6 A shop specialising in household kitchen appliances offers a delivery
service to customers. On their website it states that a 25-mile trip will
only cost $40 delivery.
The cost of delivery is directly proportional to the distance travelled.
A customer buys a cooker for $320 and lives 45 miles from the shop.
Calculate the total cost of buying the cooker.

LET'S TALK
How many
decimal places
should these
answers be given
to?

LET'S TALK
Usually larger
packs are
proportionally
cheaper than
smaller ones.

Why might this be
the case?

Inverse proportion

Sometimes an increase in one quantity causes a decrease in another quantity. For example, if bricks are being laid, then the more bricklayers there are, the less time it will take. The time taken is said to be **inversely proportional** to the number of bricklayers.

LET'S TALK

What assumptions must be made in order to answer this question?

> ### Worked examples
>
> 1 If eight fruit pickers can pick the apples from a large orchard in six days, how long will it take twelve fruit pickers to do the same job?
>
> If eight fruit pickers take six days,
>
> one fruit picker would take 6×8 days.
>
> Total 48 days.
>
> Therefore, 12 fruit pickers would take $\frac{6 \times 8}{12} = 4$ days.
>
> An alternative method would be to present the information in a table:
>
> | Number of fruit pickers | 8 | 12 |
> |---|---|---|
> | Number of days | 6 | x |
>
> For variables that are inversely proportional, their product is always constant.
>
> In this example, number of fruit pickers × number of days is constant, i.e. 48.
>
> Therefore $12x = 48$, which gives $x = 4$.
>
> 2 A cyclist averages a speed of 27 km/h for 4 hours.
> At what speed would she need to cycle to cover the same distance in 3 hours?
>
> Completing it in one hour would require cycling at 27×4 km/h.
>
> Completing it in three hours therefore requires cycling at
>
> $\frac{27 \times 4}{3} = 36$ km/h.
>
> Alternatively, place the variables in a table:
>
> | Time | 4 | 3 |
> |---|---|---|
> | Average speed | 27 | x |
>
> As the product of the two variables is constant, $4 \times 27 = 3x$.
>
> Therefore, $x = \frac{4 \times 27}{3} = 36$ km/h.

Exercise 29.4

1. A teacher shares sweets among eight students so that they get three each.

   How many sweets would they each have got if there had been twelve students?

2. Six people can dig a trench in eight hours.

   a How long would it take
      i) 4 people
      ii) 12 people
      iii) 1 person?

   b How many people would it take to dig the trench in
      i) 3 hours
      ii) 16 hours
      iii) 1 hour?

3. The table below represents the relationship between the average speed and the time taken for a train to travel between two stations.

Speed (km/h)	60			120	90	50	10
Time (h)	2	3	4				

   a Copy and complete the table.

   b If the fastest time the train has travelled the distance between the two stations is 50 minutes, calculate the average speed it was travelling at.

4. a A swimming pool is filled in 30 hours by two identical pumps. How much quicker would it be filled if five similar pumps were used instead?

   b The swimming pool owner wants it to be filled in 2 hours and 40 minutes. What is the minimum number of pumps he will need?

Now you have completed Unit 29, you may like to try the Unit 29 online knowledge test if you are using the Boost eBook.

Compound measures and graphs

- Read, draw and interpret graphs and use compound measures to compare graphs.

Travel graphs

You will already be familiar from the work covered in Stages 7 and 8 that graphs can be used to display the movement of an object over time.

The distance–time graph below shows the movement of a brother and sister over time. They leave the house at different times and move in the same direction along the same straight path. The brother goes for a gentle walk, whilst the sister goes for a jog.

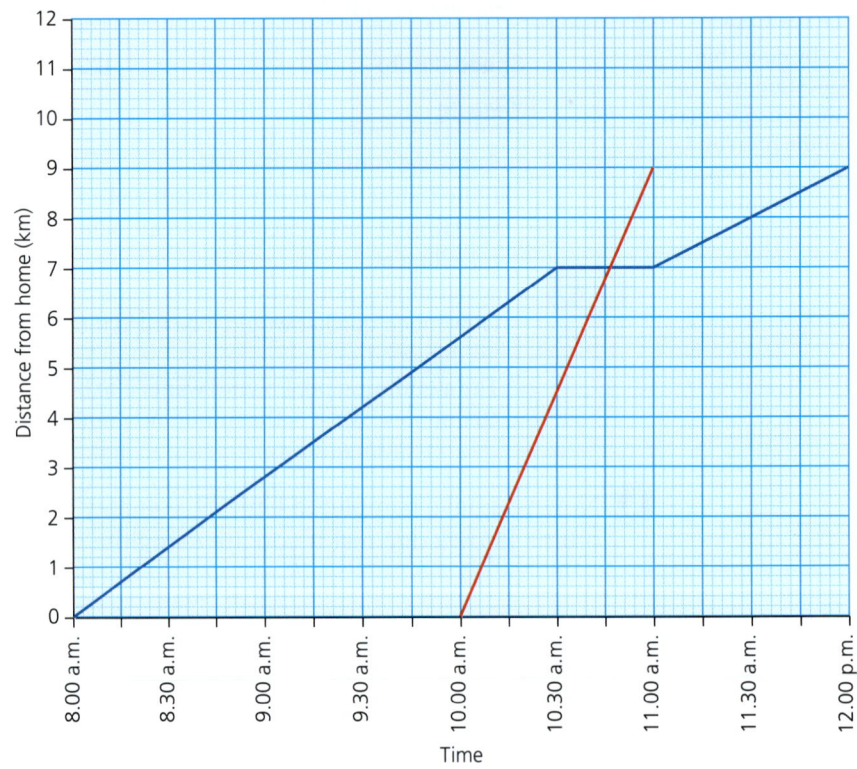

We can deduce a number of things about their movement from the graph.
- One of them leaves the house at 8.00 a.m. and the other at 10.00 a.m.
- They both end up 9 km from the house, but arrive at different times.
- The steeper line is likely to be the sister jogging as it shows that the person has covered the same distance of 9 km in less time. We will assume that this is the case.
- The sister overtakes the brother at just after 10.45 a.m. This can be seen from where the two lines cross.
- The brother stops walking between 10.30 and 11.00 a.m. This can be deduced from the fact that the line is horizontal at that point, meaning no distance was travelled.
- The speed of their movement can be calculated from the gradient of the line.

LET'S TALK

Why can we not be certain that the steeper line is the sister jogging?

Brother:
- Between 8.00 and 10.30 am he walks 7 km in 2.5 hours, so his speed is $\frac{7}{2.5} = 2.8$ km/h.
- Between 10.30 and 11.00 am he walks 0 km in 0.5 hours, so his speed is $\frac{0}{0.5} = 0$ km/h.
- Between 11.00 am and 12.00 pm he walks 2 km in 1 hour, so his speed is $\frac{2}{1} = 2$ km/h.

This shows that the gradient of a distance–time graph gives us the speed of travel.

He is therefore walking quickest when the gradient of the line is steepest.

Sister:
- Between 10.00 and 11.00 am she jogs 9 km in 1 hour, so her speed is $\frac{9}{1} = 9$ km/h.

However, not all travel graphs are distance–time graphs. It is also possible to show the motion using a **speed–time graph**.

The motion of the brother walking in the distance–time graph above could also be shown as following the a speed–time graph.

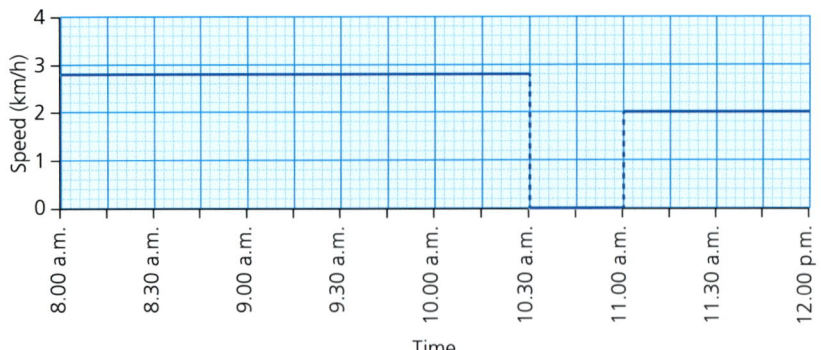

LET'S TALK

Why is this not a realistic description of someone's speed over time?

KEY INFORMATION

The horizontal lines show that he was walking at a constant speed. The dashed vertical lines simply show that one horizontal line is connected to the next.

Worked example

The distance–time graph below shows the motion of a cyclist travelling along a straight road. The distance is measured from a fixed point X on the road.

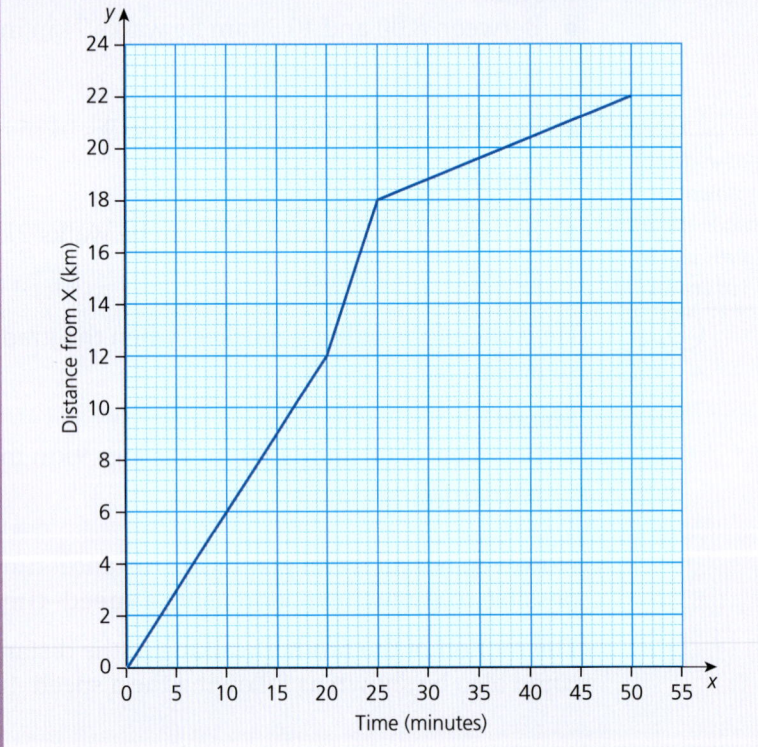

a How far away from the starting point was the cyclist after 25 minutes?

From the graph we can see that the cyclist was 18 km from the start after 25 minutes.

Note that the speed here is given as km/min because the time on the x-axis of the graph is in minutes.

LET'S TALK

How do you convert km/min to km/h?

b i) How, from the shape of the graph, can you tell when the cyclist was travelling fastest?

The steepest part of a distance–time graph shows when he is cycling fastest. This happens between 20 and 25 minutes.

ii) Calculate his top cycling speed.

The gradient of the straight line can be worked out using the coordinates of the two end points of that line segment, i.e. (20, 12) and (25, 18).

$$\text{Gradient} = \frac{y_2 - y_1}{x_2 - x_1} = \frac{18 - 12}{25 - 20} = \frac{6}{5}$$

Therefore, the speed is $\frac{6}{5} = 1.2$ km/min.

Exercise 30.1

1 This graph shows the speed of a car during a journey.

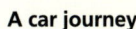

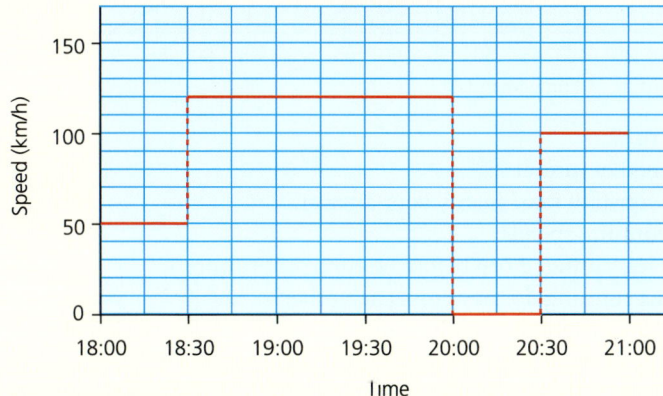

a What time did the car set off on its journey?
b How fast did the car travel for the first 30 minutes?
c For how long did the car travel at 120 km/h?
d The driver stopped for a snack along the route. At what time did this occur?
e How far did the car travel during the first half-hour?
f How far did the car travel on the whole journey?

2 Two runners run towards each other along a long straight road.
A distance–time graph of their movement is shown below.

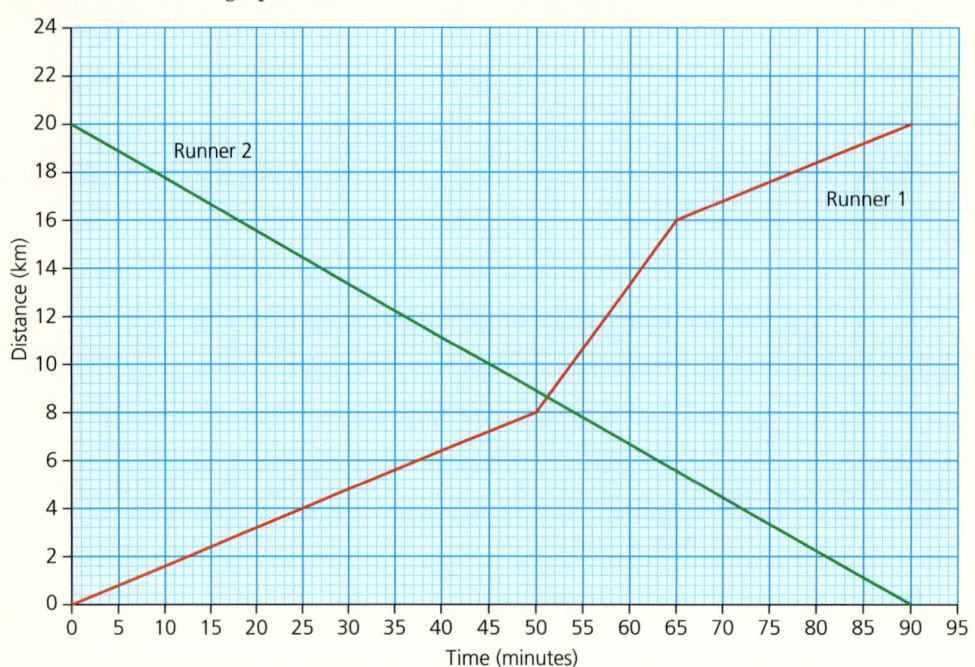

a How far does each runner run?
b i) What is the running speed of runner 2?
   ii) Does runner 2 run at a constant rate? Justify your answer.
c What is runner 1's fastest running speed?
d i) Approximately how far has runner 2 run before he runs past runner 1?
   ii) How long after they start running do they cross each other?
e State two reasons why this distance–time graph cannot be a realistic description of their movement.

3 Two walkers leave at different times of the day on the same route. This graph shows their journeys.

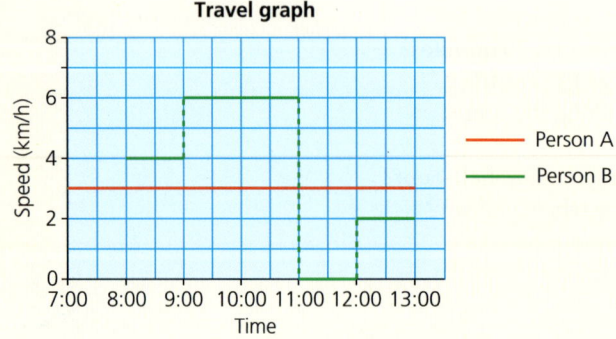

a What time did person A set off?
b At what speed did person A walk?
c For how long did person A walk?
d How far did person A walk?
e What time did person B set off?

f At what speed did person B walk between 9:00 and 11:00?
g How far did person B walk between 11:00 and 12:00?
h Who had walked further by 11:00? Show your method clearly.

4 A car sets off along a straight road. Its motion is plotted on the distance–time graph shown below.

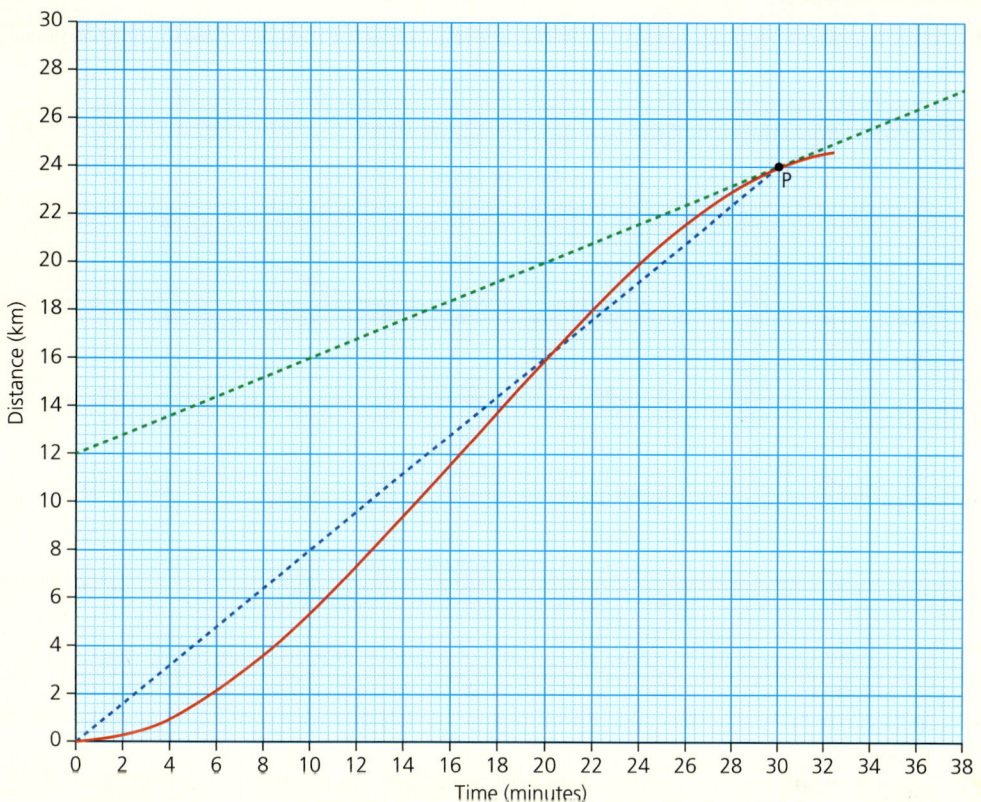

Two students, Cristina and João, are analysing the graph. They want to estimate the car's speed at a point P, 30 minutes after it sets off.
Cristina says that the speed at P can be calculated by drawing a straight line from the origin to point P and working out its gradient.
João argues that a more accurate estimate would be to work out the gradient of the tangent to the curve at point P.

a Which student is correct? Justify your answer.
b Calculate the speed of the car after 30 minutes. Give your answer in
   i) km/min   ii) km/h.
c Estimate the speed of the car after 6 minutes. Give your answer in km/h.
d Approximately when was the car travelling fastest? Justify your answer.

LET'S TALK

How can you tell if the speed of the car is constant or not from the shape of the graph?

Graphs in real-life contexts

Graphs can be used to represent many different real-life situations. The following exercise introduces a few examples of where graphs can be used in real life.

Exercise 30.2

1 Two mobile phone companies are advertising their tariffs for the cost of sending text messages.
Company A: Unlimited texts for $10 per month
Company B: 100 texts cost $1.50
This graph shows the cost of the two tariffs.

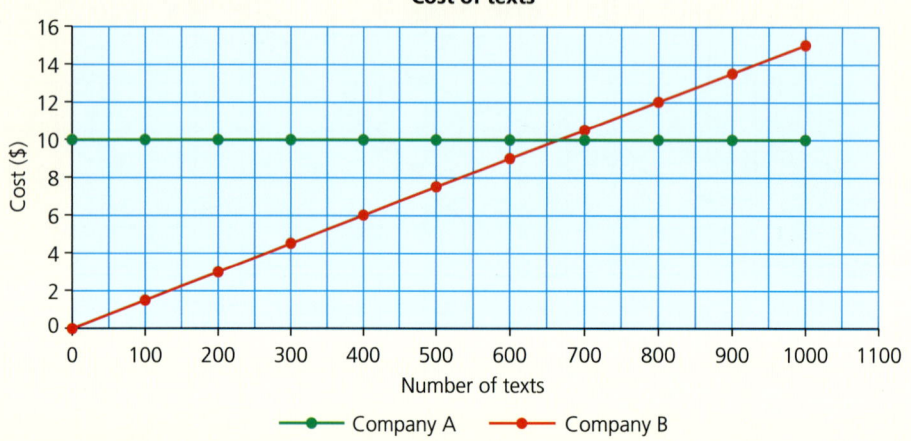

Cost of texts

a Which company is better value for money for a customer who sends 500 texts per month on average?
b Which company is better value for money for a customer who sends 1000 texts per month on average?
c Approximately how many texts does a customer need to send each month for the two companies to cost the same? Justify your answer.

2 Two new cars are bought at the same time. Car X costs \$45 000 and car Y costs \$20 000.
This graph shows their value over the next ten years.

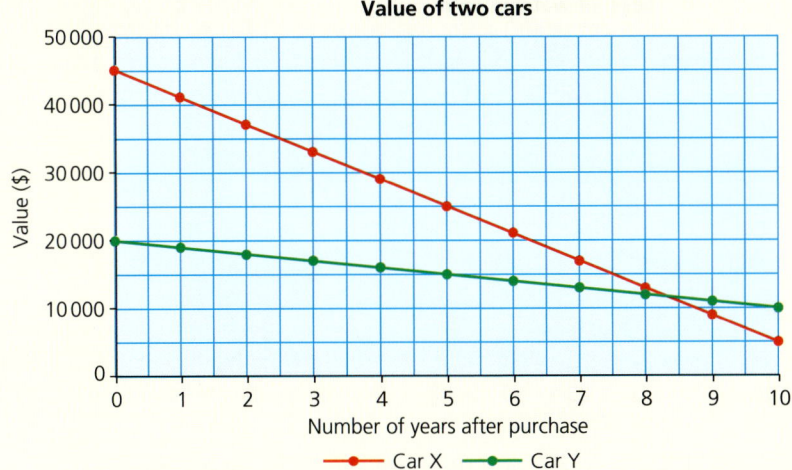

Value of two cars

a Which car is better value for money for a buyer who wants a car which does not depreciate (lose its value) quickly?

b What is the rate of depreciation for each car?

c How can you tell which car depreciates more quickly from the shape of the graph?

d What is the approximate difference in the two cars' value after five years?
Show your method clearly.

e Which car is worth more after ten years?

f Approximately how many years after they are bought are the two cars worth the same?

3 A gardener is using a watering can filled from a tank to water his garden.
This graph shows the water level (in centimetres) in the tank over a one-hour period.

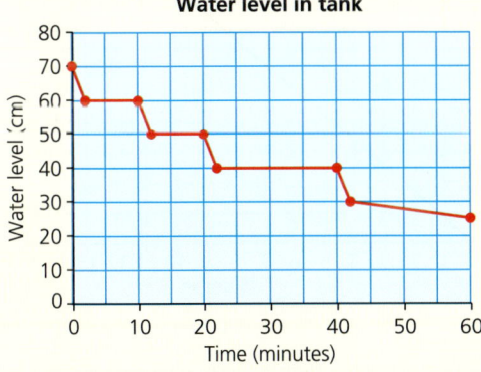

Water level in tank

a What was the level of the water in the tank at the start?

b Assuming that the gardener fills the watering can at the same rate each time, how many times does he fill the watering can?

c The gardener finishes watering his garden and leaves after 50 minutes.
Give one possible reason why the level in the water tank continues to drop.

4 A boy decides to have a bath.
This graph shows the water level (in centimetres) in the bath over time.

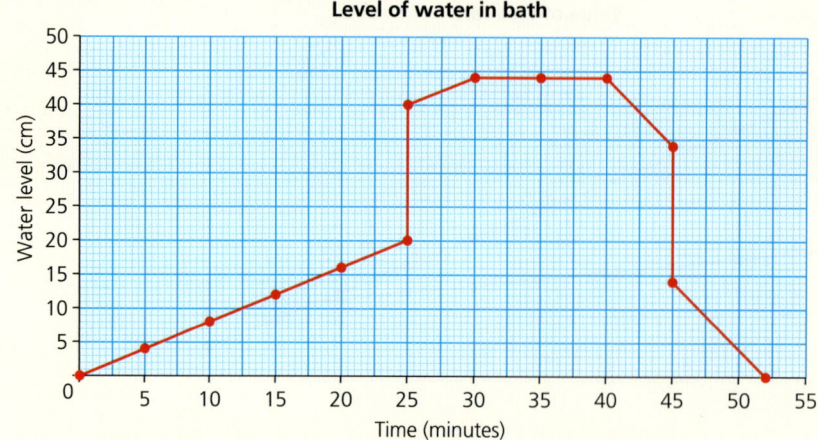

Level of water in bath

a Give a possible explanation for the shape of the graph during the first 25 minutes.
b When did the boy get into the bath?
   Justify your answer by referring to the graph.
c How long did the boy stay in the bath?
d Give a possible explanation for the shape of the graph after 45 minutes.

5 This graph shows the exchange rate between US dollars and two other currencies, the euro and the Chinese yuan.

a How many euros is equivalent to $520?

b A traveller wishes to exchange 450 euros into Chinese yuan. How many yuan can the traveller expect to receive?

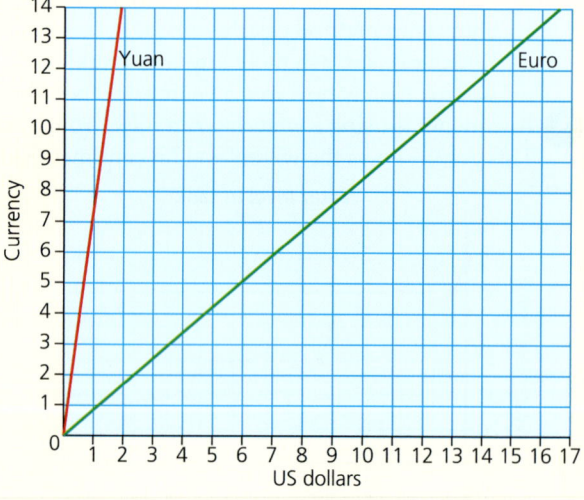

"The definition of a good mathematical problem is the mathematics it leads to rather than the problem itself."

Andrew Wiles

 Now you have completed Unit 30, you may like to try the Unit 30 online knowledge test if you are using the Boost eBook.

Section 3 – Review

1 The patterns of grey and white squares form a sequence.

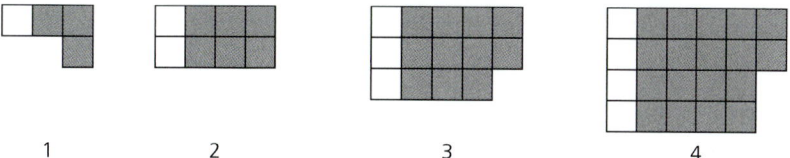

1   2   3   4

a Draw the next pattern in the sequence.

b Copy and complete the table below for the number of grey squares in each pattern.

Position	1	2	3	4	5	6
Number of grey squares						

c Write down the rule for the nth term of the number of grey squares in the sequence.

d Explain, with reference to the patterns themselves, why the rule works.

e Write down the number of white squares and the number of grey squares in the 15th pattern.

2 A sports shop usually sells a particular pair of trainers for $120. In the sale, they reduce the price of the trainers by 30%.

a Calculate the sale price.

b After the sale is over, the price is returned to $120. What is the percentage increase from the sale price?

3 Two right-angled triangles A and B are shown below. The length of one side and the area are given for each of them.

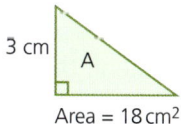

3 cm A

Area = 18 cm²

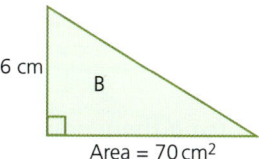

6 cm B

Area = 70 cm²

Give a reasoned explanation as to whether triangle B is an enlargement of A.

4 Write each of the following equations of a straight line in the form $ax + by = c$, where a, b and c are integer values.

a $y = -2x + 5$

b $y = \dfrac{3}{8}x - 1$

c $y = -\dfrac{4}{9}x + \dfrac{2}{3}$

5 The coordinates of a point A on a line segment AB are (5, 5) as shown.

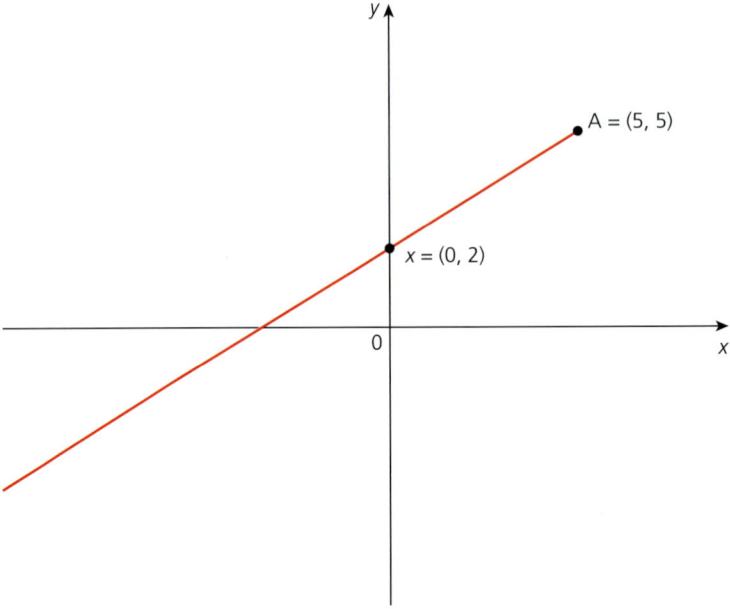

X lies on the line and has coordinates (0, 2).

If X divides AB in the ratio $2:5$, calculate the coordinates of B.

6 The volume of a cube is $850\,\text{cm}^3$.

Prove, using a reasoned argument, that the length x of each edge of the cube must lie in the range $9 < x < 10$.

7 The equations of four straight lines are given below.

$$y = \frac{2}{3}x - 4 \qquad\qquad y + \frac{2}{3}x + 2 = 0$$

$$3y + 6 = -2x \qquad\qquad 3y - 2x + 12 = 0$$

Without plotting them, identify which lines are parallel to each other. Justify your choices clearly.

8 Two villages A and B are situated 10 km apart. Village A is due
  north of B.
  Village C is on a bearing of 250° from A and a bearing of 315°
  from B.
  a Draw a scale diagram to show the positions of A, B and C
     relative to each other.
  b From your diagram estimate the distances AC and BC.
9 A hosepipe has a circular cross-section with a radius of 1 cm.
  When connected to a tap it can fill a 6000-litre container in
  4 hours.
  A different larger hosepipe has a cross-section of radius 2 cm.
  How long will it take to fill the 6000-litre container if two of the
  larger hosepipes are each connected to a tap?
10 In the year 2000, two houses M and N in different parts of the
   country have different values.
   M has a value of $400 000, whilst N has a value of $260 000.
   Over the next 20 years the value of each house changes at a
   constant rate.
   In 2020, house M is worth $320 000, whilst N is worth $305 000.
   a On the same axes, plot a graph to show the change of their
      values over the 20 years.
   b i) Calculate the annual *rate* of depreciation for house M.
      ii) How does the rate of depreciation relate to the graph?
      iii) Calculate the annual rate of appreciation for house N.
   c Assuming that the values continue to change at this constant
      rate, use your graph to estimate when the two houses would
      have the same value.

Glossary 🔊

Arc Part of a circumference of a circle.

Area factor of enlargement When an object is enlarged by a scale factor, the area also increases. The area factor is the multiple by which the area has increased. The area factor of enlargement is the scale factor of enlargement squared.

Base number In 5^3, the 5 is known as the base number. Here the base number has been raised to the power of 3.

Bias This occurs if there is prejudice either for or against a particular outcome, e.g. If questions in a questionnaire show bias then they will affect the results of the questionnaire, and if a dice is biased towards six, then it will land on six more often than it should.

Bisect To divide in half.

Byte A unit of memory size in computers.

Capacity A measure of volume dealing with liquids.

Causal effect The amount of influence one outcome has on the outcome of another event. For example, the amount of sunshine will have a causal effect on the number of ice-creams sold.

Centre of enlargement When an object is enlarged all the distances from the centre of enlargement are multiplied by the scale factor.

Coefficient This is the number (a constant) that multiples the variable in an algebraic term, for example in the $5x$ is the coefficient of the variable x.

Common difference If the numbers in a sequence change by the same amount from one term to the next, they will have a common difference. For example, in the sequence 3, 8, 13, 18, 23 etc., the common difference is +5 as the terms increase by 5 each time.

Complementary events If the probability of an event happening is $\frac{1}{3}$, the complementary event is the probability of it not happening, i.e. $\frac{2}{3}$.

Compound percentage With a compound percentage you work out the percentage of a value first then add it to the total. The percentage of that total is then calculated and added on again, and so on.

Congruent shapes Shapes that are identical in shape and size.

Cross-sectional area When a 3D shape is cut through, the shape is known as its cross-section. The area of the cross-section is known as the cross-sectional area.

Cumulative An increasing quantity formed by continued additions.

Cumulative frequency The running total of the frequencies.

Depreciation The amount the monetary value of something has decreased by.

Difference of two squares The subtraction of two square numbers, for example $5^2 - 3^2$.

Direct proportion Two variables are in direct proportion if the ratio of their values is constant.

Distributive law The distributive law says that multiplying a number by a group of numbers added together is the same as multiplying each first and then adding them together, for example $4 \times (6 + 2) = 4 \times 6 + 4 \times 2$.

Expanded form Multiplying out brackets to expand an expression, for example $3a + 12$ is the expanded form of $3(a + 4)$.

Expected frequency The number of times an event is likely to happen. For example, when a coin is flipped 100 times, the expected frequency of the number of heads is 50.

Experimental probability The probability of an event happening based on how many times it has already occurred. For example, the probability of a drawing pin landing point up can be based on the results of dropping a pin 100 times and seeing how many times it landed point up.

Exterior angle If an edge of a polygon is extended, the angle formed between the extended line and the next edge is the exterior angle, for example:

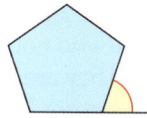

Factorised form Writing an expression with the common factors of the terms written outside the brackets, for example $3(a + 4)$ is the factorised form of $a + 12$. It is the opposite to expanded form.

Geometric proof A proof in geometry, such as proving that the three angles of a triangle add up to 180°.

Gigabyte A unit of computer memory size equivalent to a billion bytes of digital information.

Independent events Those whose outcomes do not affect each other. For example, if a dice is rolled twice, the outcome of the 1st roll will not affect the outcome of the 2nd roll.

Index The index of a number refers to the power to which it is raised.

Indices The plural of index; refers to the power to which a base number is raised.

Inscribed polygon A polygon drawn inside a circle in which all the vertices of the polygon lie on the circumference of that circle.

Interior angle The angle formed between two adjacent edges of a polygon.

Inverse proportion Inverse proportion occurs when the value of one variable decreases as the value of another variable increases, so that the product of the two variables remains constant.

Irrational number A number that cannot be written as a fraction, for example π.

Light year The distance light travels in one year.

Line of best fit A straight line drawn through a set of points so that it is as close to them all as possible.

Linear function This is the same as the equation of a straight line.

Linear sequence A sequence that increases or decreases by the same number each time, i.e. the difference between one term and the next is constant.

Lower limit The lowest possible value a number could have taken when it is approximated to a given number of significant figures or decimal places.

Megabyte A unit of computer memory size equivalent to 1 million bytes of digital information.

Microgram A unit of mass equivalent to one millionth of a gram.

Microlitre A unit of liquid capacity equivalent to one millionth of a litre.

Micrometre A unit of length equivalent to one millionth of a metre.

Mid-interval value For grouped data, it is the value represented by the middle of the group.

Milligram A unit of mass equivalent to one thousandth of a gram.

Modal class interval For grouped data, it is the group with the greatest frequency.

Mutually exclusive events Events that cannot happen at the same time. For example, when rolling a dice, getting a number that is both odd and even is impossible as they are mutually exclusive.

Nanometre A unit of length equivalent to one billionth of a metre.

Negative index A power that is negative, for example has an index that is 3^{-4}.

Perfect square A number or expression that is the square of another number or expression, for example $x^2 + 2x + 1$ is a perfect square as it is equivalent to $(x + 1)^2$.

Periodic A function or sequence that repeats itself.

Perpendicular bisector A line that cuts another line in half and meets it at right angles.

Plane of symmetry Similar to a line of reflection in 2D shapes. It divides a 3D shape into two congruent solid shapes.

Primary data Data that is directly collected by you.

Proof by contradiction A proof that starts with an assumption that something is true and then it is shown to be false (or vice versa).

Quadratic sequence A sequence of numbers where the rule for the nth term is a quadratic. Its second row of differences will be constant.

Qualitative data Any type of data that are not numerical.

Quantitative data Data that are in number form.

Regular polygon A polygon (a 2D shape where all sides are straight lines) where all sides are equal in length and all angles are the same size. A square is an example of a regular polygon.

Relative frequency The number of times an event happens divided by the total number of outcomes.

Sample size The amount of data selected to represent a whole population, for example, in a class of 30 students a sample could be 5 students.

Scale factor of enlargement How much each side of the original shape has been multiplied by to produce the enlarged shape.

Scientific notation This is the same as writing a number in standard form.

Secondary data Data that have been collected by someone else.

Similar Shapes that are mathematically similar are the same shape but a different size to each other. Their angles will be the same as each other and the lengths of their sides will stay in proportion.

Simultaneous equations When two equations are solved together in order to find the values of two variables.

Speed–time graph A type of motion graph in which the speed of something is shown over a period of time.

Spurious correlation When two variables appear to be correlated with each other but are in fact each correlated to a third variable. For example, ice-cream sales may increase as sales of jumpers decrease, but ice-cream sales do not cause the change in sales of jumpers. They are likely both correlated to outside temperature.

Standard form Numbers written in the form $a \times 10^n$ where $1 \leqslant a < 10$.

Straight-line segment Part of a straight line between two points.

Subject of the formula The single variable written on one side of a formula equal to everything else.

Surd An irrational number involving square roots, such as $\sqrt{5}$.

Terabyte A unit of computer memory size equivalent to 1 trillion (a million million) bytes of digital information.

Theoretical probability The likelihood of an event happening in theory. For example, the probability of a normal dice landing on 3 is $\frac{1}{6}$.

Tonne A unit of mass equivalent to one thousand kilograms.

Trial and improvement The process of using an answer to produce a better answer.

Upper limit The largest possible value a number could have taken when it is approximated to a given number of significant figures or decimal places.

Zero index When the power of a number is zero, for example 5^0 has zero index.

Index